# Applied Mathematical Sciences | Volume 59

# Applied Mathematical Sciences


1. John: **Partial Differential Equations**, 4th ed.
2. Sirovich: **Techniques of Asymptotic Analysis.**
3. Hale: **Theory of Functional Differential Equations**, 2nd ed.
4. Percus: **Combinatorial Methods.**
5. von Mises/Friedrichs: **Fluid Dynamics.**
6. Freiberger/Grenander: **A Short Course in Computational Probability and Statistics.**
7. Pipkin: **Lectures on Viscoelasticity Theory.**
9. Friedrichs: **Spectral Theory of Operators in Hilbert Space.**
11. Wolovich: **Linear Multivariable Systems.**
12. Berkovitz: **Optimal Control Theory.**
13. Bluman/Cole: **Similarity Methods for Differential Equations.**
14. Yoshizawa: **Stability Theory and the Existence of Periodic Solutions and Almost Periodic Solutions.**
15. Braun: **Differential Equations and Their Applications**, 3rd ed.
16. Lefschetz: **Applications of Algebraic Topology.**
17. Collatz/Wetterling: **Optimization Problems.**
18. Grenander: **Pattern Synthesis: Lectures in Pattern Theory, Vol I.**
20. Driver: **Ordinary and Delay Differential Equations.**
21. Courant/Friedrichs: **Supersonic Flow and Shock Waves.**
22. Rouche/Habets/Laloy: **Stability Theory by Liapunov's Direct Method.**
23. Lamperti: **Stochastic Processes: A Survey of the Mathematical Theory.**
24. Grenander: **Pattern Analysis: Lectures in Pattern Theory, Vol. II.**
25. Davies: **Integral Transforms and Their Applications**, 2nd ed.
26. Kushner/Clark: **Stochastic Approximation Methods for Constrained and Unconstrained Systems.**
27. de Boor: **A Practical Guide to Splines.**
28. Keilson: **Markov Chain Models—Rarity and Exponentiality.**
29. de Veubeke: **A Course in Elasticity.**
30. Sniatycki: **Geometric Quantization and Quantum Mechanics.**
31. Reid: **Sturmian Theory for Ordinary Differential Equations.**
32. Meis/Markowitz: **Numerical Solution of Partial Differential Equations.**
33. Grenander: **Regular Structures: Lectures in Pattern Theory, Vol. III.**
34. Kevorkian/Cole: **Perturbation Methods in Applied Mathematics.**
35. Carr: **Applications of Centre Manifold Theory.**
36. Bengtsson/Ghil/Källén: **Dynamic Meterology: Data Assimilation Methods.**
37. Saperstone: **Semidynamical Systems in Infinite Dimensional Spaces.**
38. Lichtenberg/Lieberman: **Regular and Stochastic Motion.**


*(continued on inside back cover)*

J.A. Sanders
F. Verhulst

# Averaging Methods in Nonlinear Dynamical Systems

With 31 Figures

Springer-Verlag
New York Berlin Heidelberg Tokyo

J.A. Sanders
Department of Mathematics
and Computer Science
Free University
1007 MA Amsterdam
The Netherlands

F. Verhulst
Mathematical Institute
State University of Utrecht
3508 TA Utrecht
The Netherlands

AMS Subject Classifications: 34A30, 34B05, 34C15, 34C29, 34C35

Library of Congress Cataloging-in-Publication Data
Sanders, J. A. (Jan A.)
Averaging methods in nonlinear dynamical systems.
(Applied mathematical sciences; v. 59)
Bibliography: p.
Includes index.
1. Differential equations, Nonlinear—Numerical solutions. 2. Differentiable dynamical systems.
3. Averaging method (Differential equations)
I. Verhulst, F. (Ferdinand), 1939– . II. Title.
III. Series.
QA372.S185 1985 515.3′55 85-22162



Printed and bound by Halliday Lithograph, West Hanover, Massachusetts.
Printed in the United States of America.

9 8 7 6 5 4 3 2 1

ISBN 0-387-96229-8 Springer-Verlag New York Berlin Heidelberg Tokyo
ISBN 3-540-96229-8 Springer-Verlag Berlin Heidelberg New York Tokyo

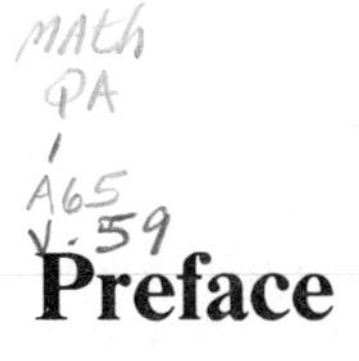


# Preface

In this book we have developed the asymptotic analysis of nonlinear dynamical systems. We have collected a large number of results, scattered throughout the literature and presented them in a way to illustrate both the underlying common theme, as well as the diversity of problems and solutions. While most of the results are known in the literature, we added new material which we hope will also be of interest to the specialists in this field.

The basic theory is discussed in chapters two and three. Improved results are obtained in chapter four in the case of stable limit sets. In chapter five we treat averaging over several angles; here the theory is less standardized, and even in our simplified approach we encounter many open problems. Chapter six deals with the definition of normal form. After making the somewhat philosophical point as to what the right definition should look like, we derive the second order normal form in the *Hamiltonian* case, using the classical method of generating functions. In chapter seven we treat *Hamiltonian* systems. The resonances in two degrees of freedom are almost completely analyzed, while we give a survey of results obtained for three degrees of freedom systems.

The appendices contain a mix of elementary results, expansions on the theory and research problems.

In order to keep the text accessible to the reader we have not formulated the theorems and proofs in their most general form, since it is our own experience that it is usually easier to generalize a simple theorem, than to apply a general one. The exception to this rule is the general averaging theory in chapter three.

Since the classic book on nonlinear oscillations by *Bogoliubov* and *Mitropolsky* appeared in the early sixties, no modern survey on averaging has been published. We hope that this book will remedy this situation and also will connect the asymptotic theory with the geometric ideas which have been so important in modern dynamics. We hope to be able to extend the scope of this book in later versions; one might e.g. think of codimension two bifurcations of vectorfields, the theory of which seems to be nearly complete now, or resonances of vectorfields, a difficult subject that one has only very recently started to research in a systematic manner.

In its original design the text would have covered both the qualitative and the quantitative theory of dynamical systems. While we were writing this text, however, several books appeared which explained the qualitative aspects better than we could ever hope to do. To have a good understanding of the geometry behind the kind of systems we are interested in, the reader is referred to the monographs of *V.I.Arnol'd* (Arn78a) , *R.Abraham* and *J.E.Marsden* (Abr78a), *J.Guckenheimer* and *Ph.Holmes* (Guc83a). A more classical part of qualitative theory, existence of periodic solutions as it is tied in with asymptotic analysis, has also been omitted as it is covered extensively in the existing literature (see e.g. (Hal69a) ).

A number of people have kindly suggested references, alterations and corrections. In particular we are indebted to *R. Cushman*, *J.J. Duistermaat*, *W. Eckhaus*, *M.A. Fekken*, *J. Schuur (MSU)*, *L. van den Broek*, *E. van der Aa*, *A.H.P. van der Burgh*, and *S.A. van Gils*. Many students provided us with lists of mathematical or typographical errors, when we used preliminary versions of the book for courses at the 'University of Utrecht', the 'Free University, Amsterdam' and at 'Michigan State University'.

We also gratefully acknowledge the generous way in which we could use the facilities of the Department of Mathematics and Computer Science of the Free University in Amsterdam, the Department of Mathematics of the University of Utrecht, and the Center for Mathematics and Computer Science in Amsterdam.

Jan A. Sanders
*Amsterdam, Summer 1985*

Ferdinand Verhulst
*Utrecht, Summer 1985*

# Contents

# 1. Basic Material

## 1.1. Introduction

In this chapter we collect some material which will play a part in the theory to be developed in the subsequent chapters. This background material consists of the existence and uniqueness theorem for initial value problems based on contraction and, associated with this, continuation results and growth estimates. Moreover we shall enumerate some theorems which arise if the system of differential equations has attraction properties.

The general form of the equations which we shall study is

$$\dot{x} = f(t,x;\epsilon),$$

where $x$ and $f$ are vectors, elements of $\mathbf{R}^n$. All quantities used will be real except if explicitly stated otherwise.

Often we shall assume $x \in D \subset \mathbf{R}^n$ with $D$ an open, bounded set. The variable $t \in \mathbf{R}$ is usually identified with time; We assume $t \geqslant 0$ or $t \geqslant t_o$ with $t_o$ a constant. The parameter $\epsilon$ plays the part of a small parameter which characterizes the magnitude of certain perturbations. We shall always take $\epsilon$ to be positive, $o < \epsilon \leqslant \epsilon_o$ with $\epsilon_o$ a constant; however, during the approximation process we may want to include the limiting value $\epsilon \downarrow 0$. We shall use $\nabla f$ to indicate the derivative with respect to the spatial variable $x$; so $\nabla f$ is the matrix with components $\dfrac{\partial f_i}{\partial x_j}$. For a vector $u \in \mathbf{R}^n$ with components $u_i, i = 1, \cdots, n$, we use the norm

$$\|u\| = \sum_{i=1}^{n} |u_i|,$$

For the $n \times n$ -matrix $A$, with elements $a_{ij}$ we have

$$\|A\| = \sum_{i,j=1}^{n} |a_{ij}|,$$

In the study of differential equations most vectors depend on variables. This involves some straightforward generalizations of concepts in calculus like

$$\int u dt,$$

which will be the vector with components

$$\int u_i(t) dt.$$

We shall not enumerate here all such generalizations which are easy to obtain. To estimate vector functions we shall nearly always use the supremum norm. For instance for the vector functions arising in the differential equation formulated above we put

$$\|f\|_{sup} = \sup_{x \in D, 0 \leqslant t \leqslant T, 0 \leqslant \epsilon \leqslant \epsilon_o} \|f(t,x;\epsilon)\|.$$

### 1.2. Existence and uniqueness of the initial value problem; continuation

The vector functions $f(t,x;\epsilon)$ arising in our study of differential equations will have certain properties with respect to the variables $t$ and $x$ and the parameter $\epsilon$. With respect to the 'spatial variable' $x$ , $f$ will always satisfy a *Lipschitz* condition:

#### 1.2.1. Definition

Consider the vector function $f(t,x;\epsilon)$ , $f \in \mathbf{R}^n$ , $t_o \leqslant t \leqslant t_o + T$ , $x \in D \subset \mathbf{R}^n$ , $0 < \epsilon \leqslant \epsilon_o$ ; $f$ satisfies a *Lipschitz* condition in $x$ with *Lipschitz* constant $L$ if in $[t_o, t_o + T] \times D \times (0, \epsilon_o]$ we have

$$\|f(t,x_1;\epsilon) - f(t,x_2;\epsilon)\| \leqslant L \|x_1 - x_2\|,$$

where $x_1, x_2 \in D$, $L$ a constant.

We are now able to formulate a well-known existence and uniqueness theorem for initial value problems.

#### 1.2.2. Theorem (existence and uniqueness)

Consider the initial value problem

$$\frac{dx}{dt} = f(t,x;\epsilon), x(t_o) = x_o$$

where $x \in D \subset \mathbf{R}^n$ , $t_o \leqslant t \leqslant t_o + T$ , $0 < \epsilon \leqslant \epsilon_o$ ; $D = \{x \mid \|x - x_o\| \leqslant d\}$ We assume that

a) $f(t,x;\epsilon)$ is continuous with respect to $t,x;\epsilon$ in $G = [t_o, t_o + T] \times D \times (0, \epsilon_o]$ .

b) $f(t,x;\epsilon)$ satisfies a *Lipschitz* condition in $x$ .

Then the initial value problem has a unique solution which exists for $t_o \leqslant t \leqslant t_o + inf(T, \frac{d}{M})$ where $M = \sup_G \|f\|$.

**Proof**

The proof of the theorem can be found in any book seriously introducing differential equations, for instance *Coddington* and *Levinson* (Cod55a) or *Roseau* (Ros66a). □

Note that the theorem guarantees the existence of a solution on an interval of time which depends explicitly on the norm of $f$. Additional assumptions enable us to prove continuation theorems, i.e. with these assumptions one can obtain existence for larger intervals or even for all time. In the sequel we shall often meet equations in the so called standard form

$$\frac{dx}{dt} = \epsilon g(t,x), x(t_o) = x_o.$$

Here, if the conditions of the existence and uniqueness theorem have been satisfied, we find that the solution exists for $t_o \leqslant t \leqslant t_o + inf(T, \frac{d}{M})$ with

$$M = \epsilon \sup_{x \in D} \sup_{t \in [t_o, t_o + T)} \|g\|.$$

Allowing $T$ to be as large as possible, this means that the size of the interval of existence of the solution is of the order $\frac{L}{\epsilon}$ with $L$ a constant. This conclusion, in which $\epsilon$ is a small parameter, involves an asymptotic estimate of the size of an interval; such estimates will be made precise in § 2.1.

## 1.3. The specific Gronwall lemma

Closely related to contraction is the idea behind an inequality derived by *Gronwall.* We formulate the inequality in a slightly different form which makes it more useful in perturbation theory.

### 1.3.1. Lemma (specific Gronwall lemma)

Suppose that for $t_o \leqslant t \leqslant t_o + T$

$$\phi(t) \leqslant \delta_2 (t - t_o) + \delta_1 \int_{t_o}^{t} \phi(s) ds + \delta_3,$$

with $\phi(t)$ continuous, $\phi(t) \geqslant 0$ for $t_o \leqslant t \leqslant t_o + T$ and constants $\delta_1 > 0$ , $\delta_2 \geqslant 0$ , $\delta_3 \geqslant 0$ then

$$\phi(t) \leqslant (\frac{\delta_2}{\delta_1} + \delta_3) e^{\delta_1 (t - t_o)} - \frac{\delta_2}{\delta_1}$$

for $t_o \leqslant t \leqslant t_o + T$ .

**Proof**

Put $\phi(t)=\psi(t)-\frac{\delta_2}{\delta_1}$ which turns the inequality into

$$\psi(t)\leqslant\delta_1\int_{t_o}^{t}\psi(s)ds + \frac{\delta_2}{\delta_1} + \delta_3.$$

Excluding the trivial case $\delta_2=\delta_3=0$, in which case $\phi(t)=0$ for $t_o\leqslant t\leqslant t_o+T$, the right hand side is positive so that

$$\frac{\delta_1\psi(t)}{\delta_1\int_{t_o}^{t}\psi(s)ds+\frac{\delta_2}{\delta_1}+\delta_3}\leqslant\delta_1.$$

Integration produces

$$\log(\delta_1\int_{t_o}^{t}\psi(s)ds + \frac{\delta_2}{\delta_1}+\delta_3) - \log(\frac{\delta_2}{\delta_1}+\delta_3)\leqslant\delta_1(t-t_o)$$

or

$$\delta_1\int_{t_o}^{t}\psi(s)ds+\frac{\delta_2}{\delta_1}+\delta_3\leqslant(\frac{\delta_2}{\delta_1}+\delta_3)e^{\delta_1(t-t_o)}.$$

Applying the original inequality for $\psi$ we have

$$\psi(t)\leqslant(\frac{\delta_2}{\delta_1}+\delta_3)e^{\delta_1(t-t_o)}$$

and transforming again to $\phi(t)$ produces the result of the lemma. □

## 1.4. Estimates in the case of attraction

Consider again the equation with initial value

$$\dot{x} = f(t,x;\epsilon),x(t_o)=x_o$$

for $t\geqslant t_o;x,x_o\in D,0<\epsilon\leqslant\epsilon_o$. Suppose that $x=0$ is a solution of the equation (if we wish to study a particular solution $x=\phi(t)$ we can always shift to an equation for $y=x-\phi(t)$, where the equation for $y$ has the trivial solution).

### 1.4.1. Definition

The solution $x=0$ of the equation is stable in the sense of *Lyapunov* if for all $\epsilon>0$ there exists a $\delta>0$ such that $\|x_o\|\leqslant\delta \Rightarrow \|x(t)\|<\epsilon$ for $t\geqslant t_o$.

The solution $x=0$ may have a different property which we call attraction:

**1.4.2. Definition**

The solution $x=0$ of the equation is a *(positive) attractor* if there is a $\delta>0$ such that

$$\|x_o\|<\delta \Rightarrow \lim_{t\to\infty} x(t) = 0.$$

If the solution is stable and moreover an attractor we have a stronger type of stability:

**1.4.3. Definition**

If the solution $x=0$ of the equation is stable in the sense of *Lyapunov* and $x=0$ is a *(positive) attractor,* the solution is *asymptotically stable.*

It is natural to study the stability characteristics of a solution by linearizing the equation in a neighborhood of this solution. One may hope that the stability characteristics of the linear equation carry over to the full non-linear equation. It turns out, however, that this is not always the case. *Poincaré* and *Lyapunov* considered some important cases where the linear behavior with respect to stability is characteristic for the full equation. In the case which we discuss, the proof is obtained by estimating explicitly the behavior of the solutions in a neighborhood of $x=0$. These explicit estimates will turn out to be useful in Ch. 4.

**1.4.4. Theorem** (*Poincaré-Lyapunov*)

Consider the equation

$$\dot{x} = (A + B(t))x + g(t,x), x(t_o)=x_o, t\geqslant t_o,$$

where $x,x_o \in \mathbf{R}^n$; $A$ is a constant $n\times n$-matrix with all eigenvalues having negative real part, $B(t)$ is a continuous $n\times n$-matrix with the property

$$\lim_{t\to\infty} \|B(t)\| = 0.$$

The vector field is continuous with respect to $t$ and $x$ and continuously differentiable with respect to $x$ in a neighborhood of $x=0$; moreover

$$g(t,x) = o(\|x\|) \text{ as } \|x\|\to 0 \text{, uniformly in } t.$$

Then there exist constants $C,t_o,\delta,\mu>0$ such that if $\|x_o\|<\dfrac{\delta}{C}$

$$\|x(t)\|\leqslant C\|x_o\|e^{-\mu(t-t_o)}, t\geqslant t_o.$$

**Remark**

The domain $\|x_o\|<\delta$ where the attraction is of exponential type will be called the *Poincaré-Lyapunov* domain of the equation.

**Proof**

Note that in a neighborhood of $x=0$, the initial value problem satisfies the conditions of the existence and uniqueness theorem.
As the matrix $A$ has eigenvalues with all real parts negative, there exists a constant $\mu_o>0$ such that for the solution of the fundamental matrix equation

$$\dot{\Phi} = A\Phi, \Phi(t_o) = I$$

we have the estimate

$$\|\Phi(t)\| \leqslant Ce^{-\mu_o(t-t_o)}, C>0, t \geqslant t_o.$$

The constant $C$ depends on $A$ only. From the assumptions on $B$ and $g$ we know that there exists $\eta(\delta)>0$ such that if $\|x\| \leqslant \delta$

$$\|B(t)\| < \eta(\delta), t \geqslant t_o(\delta),$$

$$\|g(t,x)\| \leqslant \eta(\delta)\|x\|.$$

Note that existence of the solution (in the *Poincaré-Lyapunov* domain) of the initial value problem is guaranteed on some interval $[t_o, \hat{t}]$. In the sequel we shall give estimates which show that the solution exists for all $t \geqslant t_o$. For the solution we may write the integral equation

$$x(t) = \Phi(t)x_o + \int_{t_o}^{t} \Phi(t-s+t_o)[B(s)x(s)+g(s,x(s))]ds.$$

Using the estimates for $\Phi, B$ and $g$ we have for $t \in [t_o, \hat{t}]$

$$\|x(t)\| \leqslant$$

$$\|\Phi(t)\|\|x_o\| + \int_{t_o}^{t} \|\Phi(t-s+t_o)\|[\|B(s)\|\|x(s)\| + \|g(s,x(s))\|]ds$$

$$\leqslant Ce^{-\mu_o(t-t_o)}\|x_o\| + \int_{t_o}^{t} Ce^{-\mu_o(t-s)} 2\eta\|x(s)\|ds$$

or

$$e^{\mu_o(t-t_o)}\|x(t)\| \leqslant C\|x_o\| + \int_{t_o}^{t} Ce^{\mu_o(s-t_o)} 2\eta\|x(s)\|ds.$$

Using *Gronwall*'s lemma 1.3.1 ($\delta_1 = 2C\eta$, $\delta_2 = 0, \delta_3 = C\|x_o\|$) we find

$$e^{\mu_o(t-t_o)}\|x(t)\| \leqslant C\|x_o\|e^{2C\eta(t-t_o)}$$

or

$$\|x(t)\| \leqslant C\|x_o\|e^{(2C\eta-\mu_o)(t-t_o)}.$$

Put $\mu = \mu_o - 2C\eta$; if $\delta$ (and therefore $\eta$) is small enough, $\mu$ is positive and we have

$$\|x(t)\| \leqslant C\|x_o\|e^{-\mu(t-t_o)}, t \in [t_o, \hat{t}].$$

If $\delta$ small enough, we also have

$$\|x(\hat{t})\| \leqslant \|x_o\|$$

so we can continue the estimation argument for $t \geqslant \hat{t}$; it follows that we may replace $\hat{t}$ by $\infty$ in our estimate. □

**1.4.5. Corollary:**

Under the conditions of the *Poincaré-Lyapunov* theorem, $x=0$ is asymptotically stable. The exponential attraction of the solutions is even so strong that the difference between solutions starting in a *Poincaré-Lyapunov* domain will also decrease exponentially. This is the content of the following lemma.

**1.4.6. Lemma**

Consider two solutions, $x_1(t)$ and $x_2(t)$ of the equation

$$\dot{x} = (A + B(t))x + g(t,x)$$

for which the conditions of the *Poincaré-Lyapunov* theorem have been satisfied. Starting in the *Poincaré-Lyapunov* domain we have

$$\|x_1(t)-x_2(t)\| \leqslant C\|x_1(t_o)-x_2(t_o)\|e^{-\mu(t-t_o)}$$

for $t \geqslant t_o$ and constants $C,\mu>0$.

**Proof**

Consider the equation for $y=x_1(t)-x_2(t)$

$$\dot{y} = (A + B(t))y + g(t,y + x_2(t)) - g(t,x_2(t)),$$

with initial value $y(t_o) = x_1(t_o)-x_2(t_o)$. We write the equation as

$$\dot{y} = [A + B(t) + \nabla g(t,x_2(t))]y + G(t,y),$$

with

$$G(t,y)=g(t,y+x_2(t))-g(t,x_2(t))-\nabla g(t,x_2(t))y.$$

Note that $G(t,0)=0$, $\lim_{t\to\infty} x_2(t)=0$ and as g is continuously differentiable with respect to $y$,

$$G(t,y) = o(y) \textit{ uniformly for } t \geqslant t_o.$$

It is easy to see that the equation for $y$ again satisfies the conditions of the *Poincaré-Lyapunov* theorem; only the initial time may be shifted forward by a quantity which depends on $\nabla g(t,x_2(t))$. □

**Remark**

If $t$ is large enough, we have

$$\|x_1(t) - x_2(t)\| \leqslant k \|x_1(t_o) - x_2(t_o)\|,$$

with $0 < k < 1$; we shall use this in chapter 4.2.

# 2. Asymptotics of Slow-time Processes, First Steps

## 2.1. Introduction

In this chapter we shall discuss those concepts and elementary methods in asymptotics which are necessary prerequisites for the study of slow-time processes in nonlinear oscillations. In considering a function defined by an integral or defined as the solution of a differential equation with boundary or initial conditions, approximation techniques can be useful. In the applied mathematics literature no single theory dominates but many techniques can be found based on a great variety of concepts leading in general to different results. We mention here the methods of numerical analysis, approximation by orthonormal function series in a *Hilbert* space, approximation by convergent series and the theory of asymptotic approximations. Each of these methods can be suitable to understand an explicitly given problem. In this book we consider problems where the theory of asymptotic approximations is useful and we introduce the necessary concepts in detail.
One of the first examples of an *asymptotic approximation* was discussed by *Euler* (Eul54a), or (Eul24a), who studied the series

$$\sum_{n=0}^{\infty}(-1)^n n! x^n$$

with $x \in \mathbf{R}$. This series clearly diverges for all $x \neq 0$. We shall see in a moment why *Euler* would want to study such a series in the first place, but first we remark that if $0 < x \ll 1$, the individual terms decrease in absolute value rapidly as long as $n < \frac{1}{x}$. *Euler* used the truncated series to approximate the function given by the integral

$$\int_0^\infty \frac{e^{-t}}{1+tx} dt.$$

We return to *Euler*'s example at the end of § 2.2. *Poincaré* (Chapter 8 in (Poi93a) ) gave the mathematical foundation of using a divergent series in approximating a function. In this century the theory of asymptotic approximations has expanded enormously but curiously enough only few authors concerned themselves with the foundations of the methods. Both the foundations and the applications of asymptotic analysis have been treated by *Eckhaus* (Eck79a); see also *Fraenkel* (Fra69a).

## 2.2. Concepts of asymptotic approximation

We are interested in perturbation problems of the following kind: Consider the initial value problem

$$\frac{dx}{dt} = f(t,x;\epsilon) \ , \ x(t_o)=x_o. \qquad 2.2\text{-}1$$

As usual, $t,t_o \in [0,\infty)$; $x,x_o \in \mathbf{R}^n$ and $\epsilon \in (0,\epsilon_o]$ with $\epsilon_o$ a small positive parameter. If the vectorfield $f$ is sufficiently smooth in a neighborhood of $x_o,t_o \in \mathbf{R}^n \times \mathbf{R}$ the initial value problem has a unique solution $x_\epsilon(t)$ for small values of $\epsilon$ (Cf. Theorem 1.2.2); some of the problems arising in this approximation process can be illustrated by the following examples. Consider the first order equation with initial value

$$\frac{dx}{dt} = x + \epsilon \ , \ x(0)=1.$$

The solution is $x_\epsilon(t)=(1+\epsilon)e^t - \epsilon$. We can rearrange this expression with respect to $\epsilon$:

$$x_\epsilon(t) = e^t + \epsilon(e^t - 1).$$

This result suggests that the function $e^t$ is an approximation in some sense for $x(t)$ if $t$ is not too large. In defining the concept of approximation one certainly needs a consideration of the domain of validity. A second simple example also shows that the solution does not always depend on the parameter $\epsilon$ in a smooth way:

$$\frac{dx}{dt} = -\frac{\epsilon x}{\epsilon + t} \ , \ x(0)=1.$$

The solution reads

$$x_\epsilon(t) = (\frac{\epsilon}{\epsilon+t})^\epsilon.$$

To characterize the behavior of the solution with $\epsilon$ for $t \geqslant 0$ one has to divide $\mathbf{R}^+$ into different domains. For instance, choosing $t >> \epsilon$ one can expand

$$x_\epsilon(t)=1+\epsilon\log\epsilon-\epsilon\log t+O(\frac{\epsilon}{t}).$$

Of course this expansion does not satisfy the initial condition. Such problems about the domain of validity and the form of the expansions arise in classical mechanics all the time; for some more realistic examples see (Ver75a).
To discuss these problems one has to introduce several concepts.

In ordering various terms in expansions certain functions of $\epsilon$ are useful.

**2.2.1. Definition**

A function $\delta(\epsilon)$ will be called an *order function* if $\delta(\epsilon)$ is continuous and positive (or negative) in $(0,\epsilon_o]$ and if $\lim_{\epsilon\to 0}\delta(\epsilon)$ exists. Sometimes we use subscripts like $i$ in $\delta_i(\epsilon)$ , $i=1,2,\cdots$. In many applications we shall use the set of order functions $\{\epsilon^n\}_{n=1}^\infty$; however also order functions like $\epsilon^q$ , $q\in\mathbf{Q}$ will play a part. In our second example we used the order function $\epsilon\log\epsilon$.
To compare order functions we use *Landau*'s symbols:

**2.2.2. Definition**

a) $\delta_1(\epsilon)=O(\delta_2(\epsilon))$ for $\epsilon\to 0$ if there exists a constant $k$ such that $|\delta_1(\epsilon)|\leqslant k|\delta_2(\epsilon)|$ for $\epsilon\to 0$;

b) $\delta_1(\epsilon)=o(\delta_2(\epsilon))$ for $\epsilon\to 0$ if $\lim_{\epsilon\to 0}\frac{\delta_1(\epsilon)}{\delta_2(\epsilon)}=0.$

**Examples:**

$$\epsilon^n=o(\epsilon^m) \text{ for } \epsilon\to 0 \text{ if } n>m;$$
$$\epsilon\sin(\frac{1}{\epsilon})=O(\epsilon) \text{ for } \epsilon\to 0;$$
$$\epsilon^2\log\epsilon=o(\epsilon^2\log^2\epsilon) \text{ for } \epsilon\to 0;$$
$$e^{-\frac{1}{\epsilon}}=o(\epsilon^n) \text{ for } \epsilon\to 0 \text{ and all } n\in\mathbf{N}.$$

Now $\delta_1(\epsilon)=o(\delta_2(\epsilon))$ implies $\delta_1(\epsilon)=O(\delta_2(\epsilon))$; for instance $\epsilon^2=o(\epsilon)$ and $\epsilon^2=O(\epsilon)$ as $\epsilon\to 0$. It is useful to introduce the notion of a sharp estimate of order functions:

**2.2.3. Definition**

$\delta_1(\epsilon)=O_S(\delta_2(\epsilon))$ for $\epsilon\to 0$ if $\delta_1(\epsilon)=O(\delta_2(\epsilon))$ and $\delta_1(\epsilon)\neq o(\delta_2(\epsilon))$ for $\epsilon\to 0$.

**Examples:**

$\epsilon\sin(\frac{1}{\epsilon})=O_S(\epsilon);$
$\epsilon\log\epsilon=O_S(2\epsilon\log\epsilon+\epsilon^3);$

In all problems we shall consider ordering in a neighborhood of $\epsilon=0$ so in

estimates we shall often omit 'for $\epsilon \to 0$'.
The real variable $t$ used in the initial value problem 2.2-1 will be called *time.* Extensive use shall also be made of *time-like variables* of the form $\tau = \delta(\epsilon)t$ with $\delta(\epsilon) = O(1)$. We are now able to estimate the order of magnitude of functions $\phi(t,\epsilon)$ defined in an interval $I$ , $\epsilon \in (0,\epsilon_o]$.

**Definition of order of magnitude of $\phi_\epsilon$ in $I$.**

a) $\phi_\epsilon = O(\delta(\epsilon))$ in $I$ if there exists a constant $k$ such that $\|\phi\| = O(\delta(\epsilon))$ for $\epsilon \to 0$, $\delta(\epsilon)$ an order function on $(0,\epsilon_o]$ and $\|.\|$ a norm for $\phi$ as a function of $t$;

b) $\phi_\epsilon = o(\delta(\epsilon))$ in $I$ if $\lim\limits_{\epsilon \to 0} \dfrac{\|\phi_\epsilon\|}{\delta(\epsilon)} = 0$;

c) $\phi_\epsilon = O_S(\delta(\epsilon))$ in $I$ if $\phi_\epsilon = O(\delta(\epsilon))$ and $\phi_\epsilon \neq o(\delta(\epsilon))$.

Note that this definition implies that we allow the norm of a function to be $\epsilon$-dependent. One should realize that there are different norms involved: first a norm to measure $\phi_\epsilon(t)$, then a norm, which we shall usually take to be the sup-norm, to measure $\phi_\epsilon$ (like in $\|\phi_\epsilon\| = \sup\limits_{t \in I} \|\phi_\epsilon(t)\|$) and finally one could take the supremum over $\epsilon$.
Of course, one can give the same definitions for spatial variables.

### 2.2.4. Example

We wish to estimate the order of magnitude of the error we make in approximating $\sin(t+\epsilon t)$ by $\sin(t)$ on the interval $I$. If $I$ is $[0,2\pi]$ we have for the difference of the two functions

$$\sup_{t \in [0,2\pi]} |\sin(t+\epsilon t) - \sin(t)| = O(\epsilon).$$

An additional complication is that in many problems the boundaries of the interval $I$ depend on $\epsilon$ in such a way that the interval becomes unbounded as $\epsilon$ tends to 0. For instance in the example above we might wish to compare $\sin(t+\epsilon t)$ with $\sin(t)$ on the interval $I = [0, \frac{2\pi}{\epsilon}]$. We find in the sup-norm

$$\sin(t+\epsilon t) - \sin(t) = O_S(1).$$

To estimate functions on such intervals several formulations can be found in the literature. One runs as follows. Suppose $\delta(\epsilon) = o(1)$ and we wish to estimate $\phi_\epsilon$ on $I = [0, \frac{L}{\delta(\epsilon)}]$ with $L$ a constant independent of $\epsilon$. The estimate is given as

$$\phi_\epsilon = O(\delta_o(\epsilon)) \textit{ as } \epsilon \to 0 \textit{ on } I.$$

**Abuse of notation**

In practice we shall often write this as

$$\phi_\epsilon(t)=O(\delta_o(\epsilon)) \text{ as } \epsilon\to 0 \text{ on } I.$$

For instance, the next example would already cause notational problems. In our example

$$\sin(t+\epsilon t)-\sin(t)=O(1) \text{ as } \epsilon\to 0 \text{ on } I.$$

We express such estimates often as follows:

**2.2.5. Definition of time-scale**

$\phi_\epsilon(t)=O(\delta_o(\epsilon))$ as $\epsilon\to 0$ on the *time-scale* $\delta^{-1}(\epsilon)$ if the estimate holds for $0\leqslant\delta(\epsilon)t\leqslant L$ with $L$ a constant independent of $\epsilon$. An analogous definition can be given for $o(\delta_o(\epsilon))$-estimates. Once we are able to estimate functions in terms of order functions we are able to define asymptotic approximations.

**2.2.6. Definition**

a) $\psi_\epsilon(t)$ is an asymptotic approximation of $\phi_\epsilon(t)$ on the interval $I$ if

$$\phi_\epsilon(t)-\psi_\epsilon(t)=o(1) \text{ as } \epsilon\to 0 \text{ , uniformly for } t\in I.$$

Or re-phrased for $\epsilon$-dependent intervals:

b) $\psi_\epsilon(t)$ is an asymptotic approximation of $\phi_\epsilon(t)$ on the time-scale $\delta^{-1}(\epsilon)$ if

$$\phi_\epsilon-\psi_\epsilon=o(1) \text{ as } \epsilon\to 0 \text{ on the time-scale } \delta^{-1}(\epsilon).$$

In general one obtains as approximations asymptotic series (or expansions) on some interval $I$. These are expressions of the form

$$\tilde{\phi}_\epsilon(t)=\sum_{n=1}^{m}\delta_n(\epsilon)\phi_{n_\epsilon}(t)$$

in which $\delta_n(\epsilon)$ are order functions with $\delta_{n+1}=o(\delta_n)$, $n=1,\cdots,m-1$ and for the functions $\phi_{n_\epsilon}(t)$ we have $\phi_{n_\epsilon}=\underset{\sim}{O_S}(1)$ on $I$. For a given function $\phi_\epsilon$ on the interval $i$, the asymptotic series $\phi_\epsilon$ is called a $m$-th order asymptotic approximation of $\phi_\epsilon$ on $I$ if

$$\phi_\epsilon(t)-\tilde{\phi}_\epsilon(t)=o(\delta_m(\epsilon)) \text{ on } I.$$

**2.2.7. Example**

Consider

$$\phi_\epsilon(t)=\sin(t+\epsilon t) \text{ on } I=[0,2\pi],$$

$$\tilde{\phi}_\epsilon=\sin(t)+\epsilon t\cos(t)-\tfrac{1}{2}\epsilon^2t^2\sin(t).$$

The order functions are $\delta_n(\epsilon)=\epsilon^{n-1}$, $n=1,2,3,\cdots$ and clearly

$$\phi_\epsilon(t) - \tilde{\phi}_\epsilon(t) = o(\epsilon^2) \text{ on } I,$$

so that $\tilde{\phi}_\epsilon(t)$ is a third order asymptotic approximation of $\phi_\epsilon(t)$ on $I$. Asymptotic approximations are not unique. Another third order asymptotic approximation of $\phi_\epsilon(t)$ on $I$ is

$$\psi_\epsilon(t) = \sin(t) + \epsilon\phi_{2_\epsilon}(t) - \tfrac{1}{2}\epsilon^2 t^2 \sin(t)$$

with $\phi_{2_\epsilon}(t) = \sin(\epsilon t)\cos(t)/\epsilon$. The functions $\phi_{n_\epsilon}(t)$ are not determined uniquely as is immediately clear from the definition. More serious is that for a given function different asymptotic approximations may be constructed with different sets of order functions. Consider an example given by *Eckhaus* ((Eck79a), chapter 1):

$$\phi_\epsilon(t) = (1 - \frac{\epsilon}{1+\epsilon}t)^{-1}, \quad I = [0,1].$$

One easily shows that the following expansions are asymptotic approximations of $\phi_\epsilon$ on $I$:

$$\psi_{1_\epsilon}(t) = \sum_{n=0}^{m} (\frac{\epsilon}{1+\epsilon})^n t^n,$$

$$\psi_{2_\epsilon}(t) = 1 + \sum_{n=1}^{m} \epsilon^n t(t-1)^{n-1}.$$

For the sake of completeness we return to the example discussed by *Euler* which was mentioned in the introduction. Instead of $x$ we use the variable $\epsilon \in (0, \epsilon_o]$. Basic calculus can be used to show that we may define the function $\phi_\epsilon$ by

$$\phi_\epsilon = \int_0^\infty \frac{e^{-t}}{1+\epsilon t} dt, \quad \epsilon \in (0, \epsilon_o].$$

Transform $\epsilon t = \tau$ to obtain

$$\phi_\epsilon = \frac{1}{\epsilon} \int_0^\infty \frac{e^{-\frac{\tau}{\epsilon}}}{1+\tau} d\tau$$

and by partial integration

$$\phi_\epsilon = \frac{1}{\epsilon}[-\epsilon \frac{e^{-\frac{\tau}{\epsilon}}}{1+\tau} \Big|_0^\infty - \epsilon \int_0^\infty \frac{e^{-\frac{\tau}{\epsilon}}}{(1+\tau)^2} d\tau]$$

and after repeated partial integration

$$\phi_\epsilon = 1 - \epsilon + 2\epsilon \int_0^\infty \frac{e^{-\frac{\tau}{\epsilon}}}{(1+\tau)^3} d\tau.$$

We may continue the process and define

$$\tilde{\phi}_\epsilon = \sum_{n=0}^{m} (-1)^n n! \epsilon^n.$$

It is easy to see that

$$\phi_\epsilon = \tilde{\phi}_\epsilon + R_{m_\epsilon},$$

with $R_{m_\epsilon} = (-1)^{m+1}(m+1)!\epsilon^m \int_0^\infty \frac{e^{-\frac{\tau}{\epsilon}}}{(1+\tau)^{m+2}} d\tau.$
Transforming back to $t$ we can show that

$$R_{m_\epsilon} = O(\epsilon^{m+1}),$$

so $\tilde{\phi}_\epsilon$ is an asymptotic approximation of $\phi(\epsilon)$. The expansion is in the set of order functions $\{\epsilon^n\}_{n=1}^{\infty}$ and the series is divergent.
A final remark concerns the case for which one is able to prove that an asymptotic series which has been obtained converges. This does not imply that the series converges to the function to be studied. For consider the simple example

$$\phi_\epsilon = \sin(\epsilon) + e^{-\frac{1}{\epsilon}}.$$

*Taylor* expansion of $\sin(\epsilon)$ produces the series

$$\tilde{\phi}_\epsilon = \sum_{n=0}^{m} \frac{(-1)^n \epsilon^{2n+1}}{(2n+1)!}$$

which is convergent for $m \to \infty$;
$\tilde{\phi}_\epsilon$ is an asymptotic approximation of $\phi_\epsilon$ as

$$\phi_\epsilon - \tilde{\phi}_\epsilon = O(\epsilon^{2m+3}) \textit{ for all } m \in \mathbf{N}.$$

However, the series does not converge to $\phi_\epsilon$.

In the theory of nonlinear differential equations this matter of convergence is of little practical interest. Usually the calculation of one or a few more terms in the asymptotic expansion is all that one can do within a reasonable amount of (computer) time.

## 2.3. Naive formulation of perturbation problems

We are interested in studying initial value problems of the type

$$\frac{dx}{dt} = f(t,x;\epsilon) \;,\; x(t_o) = x_o, \qquad 2.3\text{-}1$$

with $x, x_o \in D \subset \mathbf{R}^n$ , $t, t_o \in [0,\infty)$ , $\epsilon \in (0,\epsilon_o]$. The vectorfield $f$ meets the conditions of the basic existence and uniqueness theorem 1.2.2. Suppose that

$$\lim_{\epsilon \to 0} f(t,x;\epsilon) = f(t,x;0)$$

exists uniformly on $D \times I$ with $I$ a sub-interval $[t_o, A]$ of $[0,\infty)$. Then we can associate with problem 2.3-1 an 'unperturbed' problem

$$\frac{dy}{dt} = f(t,y;0) \;,\; y(t_o) = x_o \qquad 2.3\text{-}2$$

and we wish to establish the relation between the solution of problem 2.3-1 and problem 2.3-2. The relation will be expressed in terms of asymptotic approximations as introduced in § 2.1. Note that this treatment makes sense if we do not know the solution of 2.3-1 and if we can solve 2.3-2. The last assumption is not trivial as 2.3-2 is in general still non-autonomous and nonlinear.
Though this looks like a natural approach it turns out that for initial value problems the procedure in general gives poor results. We shall show this by an example and then we present a general estimate of the difference between the solutions of the perturbed and the unperturbed problem in Lemma 2.3.2.

**2.3.1. Example**

$$\frac{dx}{dt}=-\epsilon x \ , \ x(0)=1 \ ; \ x\in[0,1] \ , \ t\in[0,\infty) \ , \ \epsilon\in(0,\epsilon_o].$$

The associated unperturbed problem is

$$\frac{dy}{dt}=0 \ , \ y(0)=1.$$

So $x(t)=e^{-\epsilon t}$ , $y(t)=1$ and we have clearly no better result than

$$x_\epsilon(t)-y(t)=O(\epsilon) \textit{ on the time-scale } 1.$$

**2.3.2. Lemma**

Consider the initial value problems

$$\frac{dx}{dt}=f(t,x)+\delta(\epsilon)g(t,x;\epsilon) \ , \ x(t_o)=x_o \tag{2.3-3}$$

and

$$\frac{dy}{dt}=f(t,y) \ , \ y(t_o)=x_o, \tag{2.3-4}$$

in which $f$ and $g$ are *Lipschitz*-continuous with respect to $x$ in $D\subset\mathbf{R}^n$ and continuous with respect to $(t,x,\epsilon)\in[t_o,\infty)\times D\times(0,\epsilon_o]$; $\delta(\epsilon)$ is an order function. If $g(t,x;\epsilon)=O(1)$ on the time-scale 1 we have

$$x_\epsilon(t)-y(t)=O(\delta(\epsilon)) \textit{ on the time-scale } 1.$$

**Proof**

We write the differential equations 2.3-3 and 2.3-4 as integral equations

$$x(t)=x_o + \int_{t_o}^{t}(f(s,x(s))+\delta(\epsilon)g(s,x(s);\epsilon))ds,$$

$$y(t)=x_o + \int_{t_o}^{t} f(s,y(s);0)ds.$$

Subtracting the equations and taking the norm of the difference we have

$$\|x(t)-y(t)\| =$$

$$\| \int_{t_o}^{t} (f(s,x(s)) - f(s,y(s);0) + \delta(\epsilon) g(s,x(s);\epsilon)) ds \|$$

$$\leq \int_{t_o}^{t} \|f(s,x(s)) - f(s,y(s);0)\| ds + \delta(\epsilon) \int_{t_o}^{t} \|g(s,x(s);\epsilon)\| ds.$$

There exists a constant $M$ with $\|g(s,x;\epsilon)\| \leq M$ on $I \times D \times (0,\epsilon_o]$. The *Lipschitz*-continuity of $f$ with respect to $x$ (with constant $L$) implies more-over

$$\|x(t)-y(t)\| \leq L \int_{t_o}^{t} \|x(s)-y(s)\| ds + \delta(\epsilon) M (t-t_o).$$

We apply the *Gronwall* lemma 1.3.1 with $\delta_1(\epsilon) = L$ , $\delta_2(\epsilon) = M\delta(\epsilon)$ , $\delta_3 = 0$ to obtain

$$\|x(t)-y(t)\| \leq \delta(\epsilon) \frac{M}{L} e^{L(t-t_o)} - \delta(\epsilon) \frac{M}{L}. \qquad 2.3\text{-}5$$

We conclude from this inequality that $y$ is an asymptotic approximation of $x$ with error $\delta(\epsilon)$ if $L(t-t_o)$ is bounded by a constant independent of $\epsilon$; so the approximation is valid on the time-scale 1. Note that we have a larger time-scale, for instance $\log(\delta(\epsilon))$, if we admit larger errors, e.g. $\delta^{1/2}$. We note that if one tries to improve the accuracy by choosing an improved associated equation (by including higher order terms in $\epsilon$), the time-scale of validity is not extended. More specifically, assume that we may write

$$\frac{dx}{dt} = f(t,x) + \delta(\epsilon) g(t,x;0) + \bar{\delta}(\epsilon) h(t,x;\epsilon),$$

with $h = O(1)$ and $\bar{\delta}(\epsilon) = o(\delta(\epsilon))$. Applying the same estimation technique with $\delta_1(\epsilon) = L$ and $\delta_2(\epsilon) = \bar{\delta}(\epsilon) M$ the estimate 2.3-5 produces for $y$, the solution of

$$\frac{dy}{dt} = f(t,y) + \delta(\epsilon) g(t,y;0) \ , \ y(t_o) = x_o,$$

the following estimate for the accuracy of the approximation:

$$x(t) - y(t) = O(\bar{\delta}(\epsilon)) \text{ on the time-scale } 1.$$

To extend the time scale of validity we need more sophisticated methods. The function $y$ defined in the Lemma and in the discussion above is often called a *formal approximation* of $x$. It turns out that this formal approximation for initial value problems is in general an asymptotic approximation on the time-scale 1. Some cases where such simple formal approximations can be used to obtain asymptotic approximations on longer time-scales will be treated in chapter 4.

## 2.4. Reformulation in the standard form

We consider the perturbation problem of the form

$$\frac{dx}{dt} = f(t,x) + \epsilon g(t,x;\epsilon) \ , \ x(t_o) = x_o \tag{2.4-1}$$

and the unperturbed problem

$$\frac{dy}{dt} = f(t,y) \ , \ y(t_o) = x_o. \tag{2.4-2}$$

We assume that 2.4-2 can be solved explicitly. The solution will depend on the initial value $x_o$ and we write it as $y(t,x_o)$. So we have

$$y = y(t,z) \ , \ y(0,z) = z \ , \ z \in \mathbf{R}^n.$$

We now consider this as a transformation (method of variation of parameters or variation of constants) as follows:

$$x = y(t,z). \tag{2.4-3}$$

Using 2.4-1 and 2.4-2 we derive the differential equation for $z$

$$\frac{\partial y}{\partial t} + \frac{\partial y}{\partial z}\frac{dz}{dt} = f(t,y) + \epsilon g(t,y;\epsilon).$$

As $y$ satisfies the unperturbed equation, the first terms on the left and right cancel out. If we assume that $\frac{\partial y}{\partial z}$ is nonsingular we may write

$$\frac{dz}{dt} = \epsilon\left(\frac{\partial y(t,z)}{\partial z}\right)^{-1} g(t,y(t,z);\epsilon). \tag{2.4-4}$$

Equation 2.4-4 supplemented by the initial value of $z$ will be called a *perturbation problem in the standard form.*
In general, however, equation 2.4-4 will be unattractive. Consider for example the perturbed mathematical pendulum equation

$$\ddot{\phi} + \sin(\phi) = \epsilon g(t,\phi;\epsilon).$$

Equation 2.4-4 will in this case necessarily involve elliptic functions (Cf. appendix 8.3).

Another difficulty of a more technical nature might be that the transformation introduces nonuniformities in the time-dependent behavior, so there is no *Lipschitz* constant $L$ independent of $t$.

Still the standard form 2.4-4 may be useful to draw several general conclusions. A simple case in mathematical biology involving elementary functions is the following example.

**2.4.1. Example**

Consider two species living in a region with a restricted supply of food and a slight interaction between the species affecting their population density $x_1$ and $x_2$. We describe the population growth by the model

$$\frac{dx_1}{dt}=a_1x_1-x_1^2+\epsilon f_1(x_1,x_2)\ ,\ x_1(0)=x_1^o,$$

$$\frac{dx_2}{dt}=a_2x_2-x_2^2+\epsilon f_2(x_1,x_2)\ ,\ x_2(0)=x_2^o,$$

where the constants $a_i,x_i^o>0$ and $x_i(t)\geqslant 0$ *for* $i=1,2$. The solution of the unperturbed problem is

$$x_i(t) = \frac{a_i}{1+\frac{a_i-x_i^o}{x_i^o}e^{-t}} = \frac{a_ix_i^oe^t}{a_i+x_i^o(e^t-1)}.$$

Applying 2.4-4 we get

$$\frac{dz_i}{dt}=\epsilon e^{-t}(1+\frac{z_i}{a_i}(e^t-1))^2 f_i(.,.)\ \ z_i(0)=x_i^o\ ,\ i=1,2,$$

in which we abbreviated the expression for $f_i$.
As has been suggested earlier on, the transformation may be often unpractical, and one can see in this example why, since even if we take $f_i$ constant, the right hand side of the equation grows exponentially. There is however an important class of problems where this technique works well and we shall treat this in the next section.

## 2.5. The standard form in the quasilinear case

The perturbation problem 2.4-1 will be called *quasilinear* if the equation can be written as

$$\frac{dx}{dt}=A(t)x+\epsilon g(t,x;\epsilon)\ ,\ x(t_o)=x_o \qquad 2.5\text{-}1$$

in which $A(t)$ is a continuous $n\times n$-matrix. The unperturbed problem

$$\frac{dy}{dt}=A(t)y$$

possesses $n$ linearly independent solutions from which we construct the fundamental matrix $\Phi(t)$. We choose $\Phi$ such that $\Phi(t_o)=I$. We apply the variation of constants procedure

$$x=\Phi(t)z$$

and we obtain from 2.4-4

$$\frac{dz}{dt}=\epsilon\Phi^{-1}(t)g(t,\Phi(t)z;\epsilon). \qquad 2.5\text{-}2$$

If $A$ is a constant matrix we have for the fundamental matrix

$$\Phi(t)=e^{A(t-t_o)}.$$

The standard form becomes in this case

$$\frac{dz}{dt}=\epsilon e^{-A(t-t_o)}g(t,e^{A(t-t_o)}z;\epsilon). \qquad 2.5\text{-}3$$

Clearly if the eigenvalues of $A$ are not all purely imaginary, the perturbation equation 2.5-3 may present some serious problems even if $g$ is bounded.

**Remark**

In the theory of forced nonlinear oscillations the perturbation problem may be of the form

$$\frac{dx}{dt}=Ax+f(t)+\epsilon g(t,x;\epsilon)\ ,\ x(t_o)=x_o, \qquad 2.5\text{-}4$$

with $A$ a constant matrix. The variation of constants transformation then becomes

$$x=e^{A(t-t_o)}z+e^{A(t-t_o)}\int_{t_o}^{t}e^{-A(\tau-t_o)}f(\tau)d\tau. \qquad 2.5\text{-}5$$

The perturbation problem in the standard form is

$$\frac{dz}{dt}=\epsilon e^{-A(t-t_o)}g(t,x;\epsilon),$$

in which $x$ still has to be replaced by expression 2.5-5.

### 2.5.1. Example

In studying nonlinear oscillations one often considers the perturbed initial value problem

$$\ddot{x}+\omega^2x=\epsilon g(t,x,\dot{x};\epsilon)\ ,\ x(t_o)=\alpha\ ,\ \dot{x}(t_o)=\beta. \qquad 2.5\text{-}6$$

Two independent solutions of the unperturbed problem $\ddot{y}+\omega^2y=0$ are $\cos(\omega(t-t_o))$ and $\sin(\omega(t-t_o))$. The variation of constants transformation becomes

$$x=z_1\cos(\omega(t-t_o))+\frac{z_2}{\omega}\sin(\omega(t-t_o)), \qquad 2.5\text{-}7$$

$$\dot{x}=-z_1\omega\sin(\omega(t-t_o))+z_2\cos(\omega(t-t_o)).$$

Note that the fundamental matrix is such that $\Phi(t_o)=I$. Equation 2.5-3 becomes in this case

$$\frac{dz_1}{dt}=-\ \frac{\epsilon}{\omega}\sin(\omega(t-t_o))g(t,.,.;\epsilon)\ ,\ z_1(t_o)=\alpha, \qquad 2.5\text{-}8$$

$$\frac{dz_2}{dt}=\epsilon\cos(\omega(t-t_o))g(t,.,.;\epsilon)\ ,\quad z_2(t_o)=\beta.$$

The expressions for $x$ and $\dot{x}$ have to be substituted in $g$ on the dotted

places.
It may be useful to adopt a transformation which immediately provides us with equations for the variation of the amplitude $r$ and the phase $\psi$ of the solution. We put

$$x = r\cos(\omega t + \psi), \qquad 2.5\text{-}9$$

$$\dot{x} = -r\omega\sin(\omega t + \psi).$$

The perturbation equations become

$$\frac{dr}{dt} = -\frac{\epsilon}{\omega}\sin(\omega t + \psi)g(t,.,.;\epsilon), \qquad 2.5\text{-}10$$

$$\frac{d\psi}{dt} = -\frac{\epsilon}{r\omega}\cos(\omega t + \psi)g(t,.,.;\epsilon).$$

The initial values for $r$ and $\psi$ can be calculated from 2.5-9. It is clear that the perturbation formulation 2.5-10 may be less attractive in problems where the amplitude $r$ can become small. In § 2.7 we show the usefulness of both transformation 2.5-7 and 2.5-9.

## 2.6. Averaging, a first introduction

In this section we shall present the simplest form of the theory of averaging. We are concerned with solving a perturbation problem in the standard form which we write as

$$\frac{dx}{dt} = \epsilon f(t,x) + \epsilon^2 g(t,x,\epsilon)\ ,\ x(t_o) = x_o. \qquad 2.6\text{-}1$$

Suppose that $f$ is $T$-periodic in $t$, then it seems natural to average $f$ over $t$ (while holding $x$ constant). So we consider the averaged equation

$$\frac{dy}{dt} = \epsilon f^o(y)\ ,\ y(t_o) = x_o, \qquad 2.6\text{-}2$$

with

$$f^o(y) = \frac{1}{T}\int_0^T f(t,y)dt.$$

Making some assumptions (not all really necessary), we can establish the following asymptotic result:

### 2.6.1. Theorem ( first order averaging )

Consider the initial value problems 2.6-1 and 2.6-2 with $x,y,x_o \in D \subset \mathbf{R}^n$ , $t \in [t_o,\infty)$ , $\epsilon \in (0,\epsilon_o]$. Suppose

a) $f,g$ and $\nabla f$ are defined, continuous and bounded by a constant $M$ independent of $\epsilon$, in $[t_o,\infty) \times D$;

b) $g$ is *Lipschitz*-continuous with respect to $x \in D$;

c) $f$ is $T$-periodic in $t$ with $T$ a constant, independent of $\epsilon$;

d) $y(t)$ belongs to an ( $\epsilon$-independent ) interior subset of $D$ on the time-scale $\frac{1}{\epsilon}$;

then

$$x(t)-y(t)=O(\epsilon) \text{ as } \epsilon\to 0 \text{ on the time scale } \frac{1}{\epsilon}.$$

**2.6.2. Remark**

One can verify that in assumption (a) it suffices to consider the set $[t_o,t_o+\frac{L}{\epsilon})$ with $L$ a constant independent of $\epsilon$.

**Proof**

Assumptions (a) and (b) guarantee the existence and uniqueness of the solutions of both initial value problems 2.6-1 and 2.6.2; by scaling $\tau=\epsilon t$ we have existence of the solution of equation 2.6-2 on the time-scale 1 in $\tau$ i.e. on $\frac{1}{\epsilon}$ in $t$. Define

$$u^1(t,y)=\int_{t_o}^{t}[f(s,y)-f^o(y)]ds.$$

We have $\|u^1(t,y)\|\leqslant 2MT$ on $[t_o,\infty)\times D$. Introduce $z(t)=y(t)+\epsilon u^1(t,y(t))$. We estimate, since $y(t)\in D$ for all $t\geqslant t_o$,

$$\|x(t)-y(t)\|\leqslant\|x(t)-z(t)\|+\|z(t)-y(t)\|$$

$$\leqslant\|x(t)-z(t)\|+\epsilon\|u^1(t,y(t))\|\leqslant\|x(t)-z(t)\|+2\epsilon MT.$$

We note that $x(t)-z(t)=\int_{t_o}^{t}(\frac{dx}{dt}-\frac{dz}{dt})ds$ and calculate

$$\frac{dx}{dt}-\frac{dz}{dt}=$$

$$\epsilon f(t,x(t))+\epsilon^2 g(t,x(t))-\frac{dy}{dt}-\epsilon\nabla u^1(t,y(t))-\epsilon\frac{\partial u^1(t,y(t))}{\partial t}$$

$$=\epsilon f(t,x(t))-\epsilon f(t,z(t))+R,$$

with

$$R=\epsilon^2 g(t,x(t),\epsilon)-\epsilon^2\nabla u^1(t,y(t))f^o(y(t))+\epsilon f(t,z(t))-\epsilon f(t,y(t)).$$

We have $\|f^o(y)\|\leqslant M$ and $\|\nabla u^1(t,y)\|\leqslant 2MT$; from the *Lipschitz*-continuity of $f$ follows that

$$\|f(t,z(t))-f(t,y(t))\|\leqslant L\|z(t)-y(t)\|\leqslant\epsilon L\|u^1(t,y(t))\|\leqslant 2\epsilon LMT.$$

So there exists a constant $k$ such that

$$\|R\|\leqslant k\epsilon^2.$$

Clearly

$$\|x(t)-z(t)\| \leqslant \int_{t_o}^{t} \| \frac{dx}{dt} - \frac{dz}{dt} \| ds$$

$$\leqslant \epsilon \int_{t_o}^{t} \| f(s,x(s)) - f(s,z(s)) \| ds + k\epsilon^2 (t-t_o)$$

$$\leqslant \epsilon L \int_{t_o}^{t} \| x(s) - z(s) \| ds + k\epsilon^2 (t-t_o).$$

Applying the specific *Gronwall* lemma 1.3.1 we find

$$\|x(t)-z(t)\| \leqslant \epsilon \frac{k}{L} e^{\epsilon L(t-t_o)} - \epsilon \frac{k}{L}$$

and consequently

$$\|x(t)-y(t)\| \leqslant \epsilon \left( \frac{k}{L} e^{\epsilon L(t-t_o)} - \frac{k}{L} + 2MT \right).$$

So if $\epsilon L(t-t_o)$ is bounded by a constant independent of $\epsilon$ we have $x(t)=y(t)+O(\epsilon)$ as $\epsilon \rightarrow 0$. □

It can be useful to have a slightly different version of this theorem if one has to allow for errors in the initial value; also the order functions $\epsilon, \epsilon^2$ may have to be more general:

**2.6.3. Theorem ( first order averaging, slightly generalized )**

Consider the initial value problem

$$\frac{dx}{dt} = \delta_1(\epsilon) f(t,x) + \delta_2(\epsilon) g(t,x,\epsilon) \ , \ x(t_o) = x_o \qquad 2.6\text{-}3$$

and the associated problem

$$\frac{dy}{dt} = \delta_1(\epsilon) f^o(y) \ , \ y(t_o) = x_o + \delta_3(\epsilon). \qquad 2.6\text{-}4$$

$\delta_1, \delta_2$ and $\delta_3$ are order functions in $\epsilon$, $\delta_2 = o(\delta_1)$ as $\epsilon \rightarrow 0$. With conditions (a-d) of Theorem 2.6.1 we have

$$x(t) = y(t) + O\left(\frac{\delta_2}{\delta_1} + \delta_3\right) \text{ on the time-scale } \frac{1}{\delta_1} \text{ as } \epsilon \rightarrow 0.$$

**Proof**

Runs along precisely the same lines as the proof of Theorem 2.6.1 (Exercise!). □

## 2.7. Examples and counter-examples of elementary averaging

In this section we shall apply the averaging theorem 2.6.1 to some classical problems. For more examples see for instance *Bogoliubov* and *Mitropolsky* (Bog61a). Also we present several counter-examples to show the necessity of some of the assumptions and restrictions of the theorem.

**2.7.1. Example (*van der Pol* equation)**

Consider the equation

$$\ddot{x}+x=\epsilon f(x,\dot{x}), \qquad 2.7\text{-}1$$

with initial values $x(0)$ and $\dot{x}(0)$ given and $f$ a sufficiently smooth function in $D\subset\mathbf{R}^2$. This is a quasilinear system (§ 2.5) and we use the phase-amplitude transformation 2.5-9 to put the system in the standard form. Put

$$x=r\cos(t+\psi),$$

$$\dot{x}=-r\sin(t+\psi).$$

The perturbation equations become (2.5-10)

$$\frac{dr}{dt}=-\epsilon\sin(t+\psi)f(r\cos(t+\psi),-r\sin(t+\psi))\ ,\ r(0)=r_o, \qquad 2.7\text{-}2$$

$$\frac{d\psi}{dt}=-\frac{\epsilon}{r}\cos(t+\psi)f(r\cos(t+\psi),-r\sin(t+\psi))\ \ ,\ \psi(0)=\psi_o.$$

We note that the vectorfield is $2\pi$-periodic in $t$ and that if $f\in C^1(D)$ we may average the right hand side as long as we exclude a neighborhood of the origin. Since the original equation is autonomous, the averaged equation depends only on $r$ and we define the two components of the averaged vectorfield as follows

$$f_1(r)=\frac{1}{2\pi}\int_0^{2\pi}\sin(t+\psi)f(r\cos(t+\psi),-r\sin(t+\psi))dt$$

$$=\frac{1}{2\pi}\int_0^{2\pi}\sin(\tau)f(r\cos(\tau),-r\sin(\tau))d\tau$$

and

$$f_2(r)=\frac{1}{2\pi}\int_0^{2\pi}\cos(\tau)f(r\cos(\tau),-r\sin(\tau))d\tau.$$

An asymptotic approximation can be obtained by solving

$$\frac{d\tilde{r}}{dt}=-\epsilon f_1(\tilde{r})\ ,\ \frac{d\tilde{\psi}}{dt}=-\epsilon\frac{f_2(\tilde{r})}{\tilde{r}}$$

with appropriate initial values. This is a reduction to the problem of solving a first order autonomous system.

We specify this for a famous example, the *van der Pol* equation:

$$\ddot{x}+x=\epsilon(1-x^2)\dot{x}.$$

We find

$$\frac{d\tilde{r}}{dt}=\frac{\epsilon}{2}\tilde{r}(1-\frac{\tilde{r}^2}{4})\ ,\ \frac{d\tilde{\psi}}{dt}=0.$$

If the initial value of the amplitude $r_o$ equals 0 or 2 the amplitude $\tilde{r}$ is

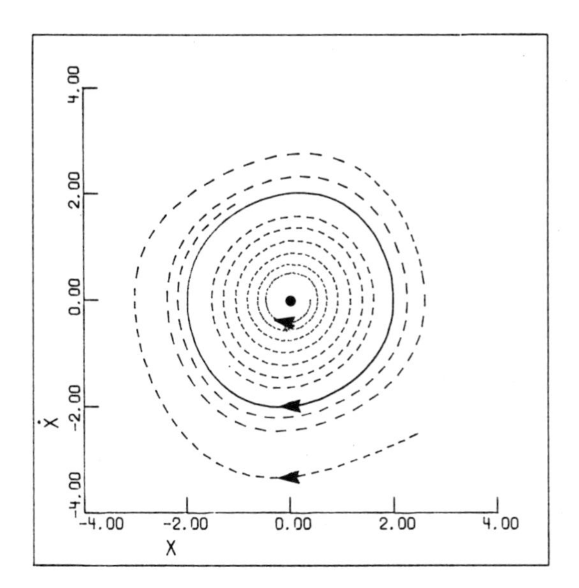


**Figure 2.7-1:** *Phase orbits of the van der Pol equation* $\ddot{x}+x=\epsilon(1-x^2)\dot{x}$ *where* $\epsilon=0.1$. *The origin is a critical point of the flow, the limit cycle (closed curve) corresponds with a stable periodic solution.*

constant for all time. $r_o=0$ corresponds with an unstable critical point of the original equation, $r_o=2$ gives a periodic solution. We have

$$x(t)=2\cos(t+\psi_o)+O(\epsilon) \text{ on the time-scale } \frac{1}{\epsilon}. \qquad 2.7\text{-}3$$

In general we find

$$x(t)=\frac{r_o e^{\frac{1}{2}\epsilon t}}{(1+\frac{1}{4}r_o^2(e^{\epsilon t}-1))^{\frac{1}{2}}}\cos(t+\psi_o)+O(\epsilon) \qquad 2.7\text{-}4$$

on the time-scale $\frac{1}{\epsilon}$. The solutions tend towards the periodic solution 2.7-3 and we call its phase orbit a (stable) limit-cycle. In figure 2.7-1 we depict some of the orbits.

In the following example we shall show that an appropriate choice of the transformation into standard form may simplify the analysis of the perturbation problem.

**2.7.2. Example (linear oscillator with frequency modulation)**

Consider an example of *Mathieu*'s equation

$$\ddot{x}+(1+2\epsilon\cos(2t))x=0,$$

with initial values $x(0)=\alpha$ and $\dot{x}(0)=\beta$.

We may proceed as in example 2.7.1; however equation 2.7-1 now explicitly depends on $t$. The phase-amplitude transformation produces with

$f = -2\cos(2t)x$

$$\frac{dr}{dt} = 2\epsilon r\sin(t+\psi)\cos(t+\psi)\cos(2t),$$

$$\frac{d\psi}{dt} = 2\epsilon\cos^2(t+\psi)\cos(2t).$$

The right hand side is $2\pi$-periodic in $t$; averaging produces

$$\frac{d\tilde{r}}{dt} = \tfrac{1}{2}\epsilon\tilde{r}\sin(2\tilde{\psi})$$

$$\frac{d\tilde{\psi}}{dt} = \tfrac{1}{2}\epsilon\cos(2\tilde{\psi})$$

So to approximate the solutions of a linear system we have to solve a non-linear system. Here the integration can be carried out but it is more practical to choose a different transformation to obtain the standard form, staying inside the category of linear systems with linear transformations. We use transformation 2.5-7 with $\omega = 1$ and $t_o = 0$:

$$x = z_1\cos(t) + z_2\sin(t),$$

$$\dot{x} = -z_1\sin(t) + z_2\cos(t).$$

The perturbation equations become (Cf. formula 2.5-8)

$$\frac{dz_1}{dt} = 2\epsilon\sin(t)\cos(2t)(z_1\cos(t) + z_2\sin(t)),$$

$$\frac{dz_2}{dt} = -2\epsilon\cos(t)\cos(2t)(z_1\cos(t) + z_2\sin(t)).$$

The right hand side is $2\pi$-periodic in $t$; averaging produces

$$\frac{d\tilde{z}_1}{dt} = \tfrac{1}{2}\epsilon\tilde{z}_2 \quad , \ \tilde{z}_1(0) = \alpha,$$

$$\frac{d\tilde{z}_2}{dt} = -\tfrac{1}{2}\epsilon\tilde{z}_1 \ , \ \tilde{z}_2(0) = \beta.$$

This is a linear system with solutions

$$\tilde{z}_1(t) = \tfrac{1}{2}(\alpha+\beta)e^{-\frac{1}{2}\epsilon t} + \tfrac{1}{2}(\alpha-\beta)e^{\frac{1}{2}\epsilon t},$$

$$\tilde{z}_2(t) = \tfrac{1}{2}(\alpha+\beta)e^{-\frac{1}{2}\epsilon t} - \tfrac{1}{2}(\alpha-\beta)e^{\frac{1}{2}\epsilon t}.$$

The asymptotic approximation for the solution $x(t)$ of this *Mathieu* equation reads

$$\tilde{x}(t) = \tfrac{1}{2}(\alpha+\beta)e^{-\frac{1}{2}\epsilon t}(\cos(t) + \sin(t)) + \tfrac{1}{2}(\alpha-\beta)e^{\frac{1}{2}\epsilon t}(\cos(t) - \sin(t)).$$

We note that the equilibrium solution $x = \dot{x} = 0$ is unstable.

In the following example an amplitude-phase representation is more appropriate.

**2.7.3. Example**

Consider the equation of motion of a one degree of freedom *Hamiltonian* system.

$$\ddot{x}+x=\epsilon f(x),$$

where $f$ is sufficiently smooth. Applying the formulae of example 2.7.1 we obtain for the amplitude and phase the following equations

$$\frac{dr}{dt}=-\epsilon\sin(t+\psi)f(r\cos(t+\psi)),$$

$$\frac{d\psi}{dt}=-\epsilon\frac{\cos(t+\psi)}{r}f(r\cos(t+\psi)).$$

We have

$$\int_0^{2\pi}\sin(t+\psi)f(r\cos(t+\psi))dt=0.$$

So the averaged equation for the amplitude is

$$\frac{d\tilde{r}}{dt}=0,$$

i.e. in first approximation the amplitude is constant. This means that a small *Hamiltonian* perturbation of the harmonic oscillator produces periodic solutions with a constant amplitude but in general a period depending on this amplitude, i.e. on the initial values.

It is easy to verify that one can obtain the same result by using transformation 2.5-7 but the calculation is much more complicated.

**2.7.4. Example (the necessity of restricting the interval of time)**

Consider the equation

$$\ddot{x}+x=8\epsilon\dot{x}^2\cos(t),$$

with initial values $x(0)=0$, $\dot{x}(0)=1$. Reduction to the standard form using the phase-amplitude transformation 2.5-9 produces

$$\frac{dr}{dt}=-8\epsilon r^2\sin^3(t+\psi)\cos(t) \qquad , r(0)=1,$$

$$\frac{d\psi}{dt}=-8\epsilon r\sin^2(t+\psi)\cos(t+\psi)\cos(t)\ ,\ \psi(0)=\ -\ \frac{\pi}{2}.$$

Averaging gives the associated system

$$\frac{d\tilde{r}}{dt}=-3\epsilon\tilde{r}^2\sin(\tilde{\psi}),$$

$$\frac{d\tilde{\psi}}{dt}=-\epsilon\tilde{r}\cos(\tilde{\psi}).$$

Integration of the system and applying the initial values yields

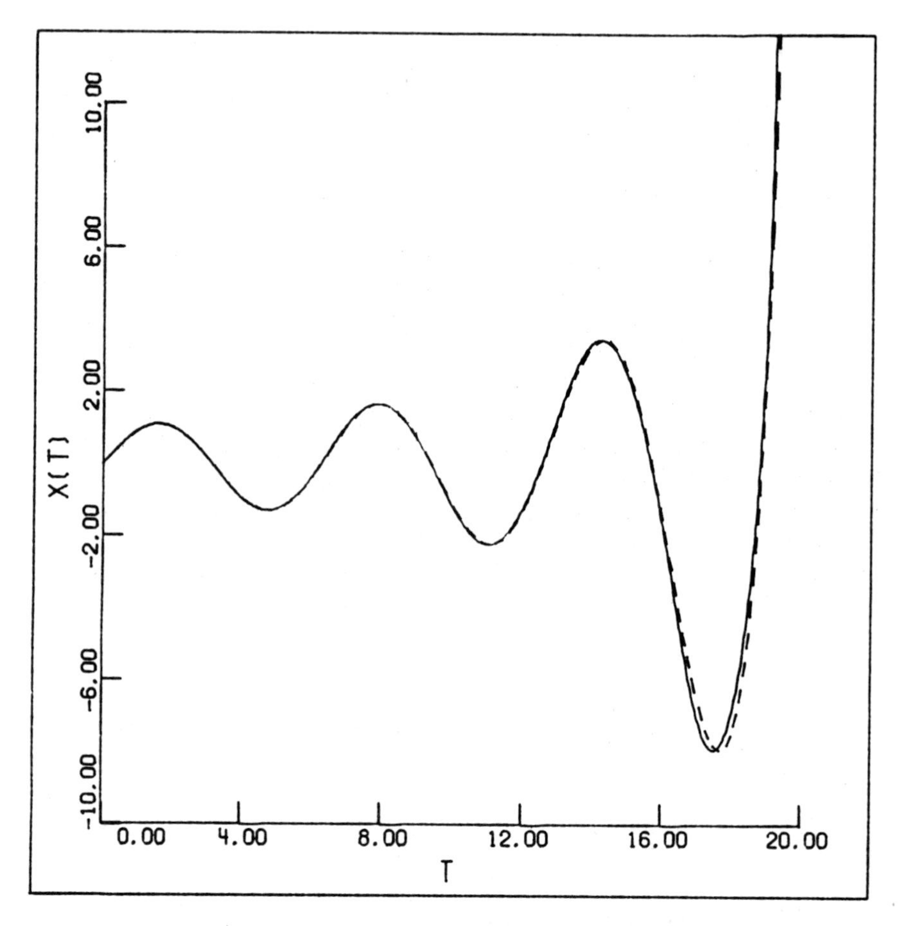


**Figure 2.7-2** *Solution $x(t)$ of $\ddot{x}+x = \frac{2}{15}\dot{x}^2\cos(t)$, $x(0)=0$, $\dot{x}(0)=1$. The solution obtained by numerical integration has been drawn full line, the asymptotic approximation has been indicated by — — —.*

$$x(t)=\frac{\sin(t)}{1-3\epsilon t}+O(\epsilon) \text{ on the time-scale } \frac{1}{\epsilon}.$$

A similar estimate holds for the derivative $\dot{x}$. The approximate solution is bounded if $0\leqslant\epsilon t\leqslant C<\frac{1}{3}$. In figure 2.7-2 we depict the approximate solution and the solution obtained by numerical integration.

One might wonder whether the necessity to restrict the time-scale is tied in with the characteristic of solutions becoming unbounded as in example 2.7.4. A simple example suffices to contradict this.

**2.7.5. Example (a trivial example with bounded solutions and a restricted time-scale of validity)**

Consider the equation

$$\ddot{x}+x=\epsilon x\ ,\ x(0)=1\ ,\ \dot{x}(0)=0.$$

After phase-amplitude transformation and averaging as in example 2.7.1 we obtain

$$\frac{d\tilde{r}}{dt}=0\quad ,\ \tilde{r}(0)=1,$$

$$\frac{d\tilde{\psi}}{dt}=-\tfrac{1}{2}\epsilon\ ,\ \tilde{\psi}(0)=0.$$

We have the approximations

$$\tilde{x}(t)=\cos((1-\tfrac{1}{2}\epsilon)t),$$

$$\dot{\tilde{x}}=-\sin((1-\tfrac{1}{2}\epsilon)t).$$

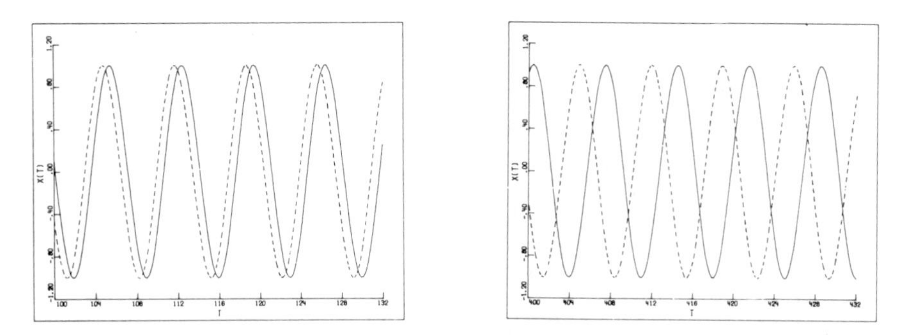

**Figure 2.7-3** *Exact and approximate solutions of* $\ddot{x}+x=\epsilon x$, $x(0)=1$, $\dot{x}(0)=0$; $\epsilon=0.1$. *The exact solution has been drawn full line, the asymptotic approximation has been indicated by* $-\ -\ -$.

Since $x(t)-\tilde{x}(t)=O(\epsilon)$ on the time-scale $\frac{1}{\epsilon}$ and

$$x(t)=\cos((1-\epsilon)^{\frac{1}{2}}t),$$

it follows that we here have an example where the approximation on $\frac{1}{\epsilon}$ is not valid on $\frac{1}{\epsilon^2}$ since obviously $x(t)-\tilde{x}(t)=O_S(1)$ on the time-scale $\frac{1}{\epsilon^2}$. In figure 2.7-3 we draw $x(t)$ and $\tilde{x}(t)$ on various time-scales.

Finally one might ask oneself why it is necessary to do the averaging after (perhaps) troublesome transformations into the standard form. Why not average small periodic terms in the original equation ? We shall call this *crude averaging* and this is a procedure that is being used by several authors. The following counter example may serve to discourage this.

**2.7.6. Example (counter example of crude averaging)**

Consider the equation

$$\ddot{x} + 4\epsilon\cos^2(t)\dot{x} + x = 0,$$

with initial conditions $x(0)=0$, $\dot{x}(0)=1$. The equation corresponds with an oscillator with linear damping where the friction coefficient oscillates between 0 and $4\epsilon$.

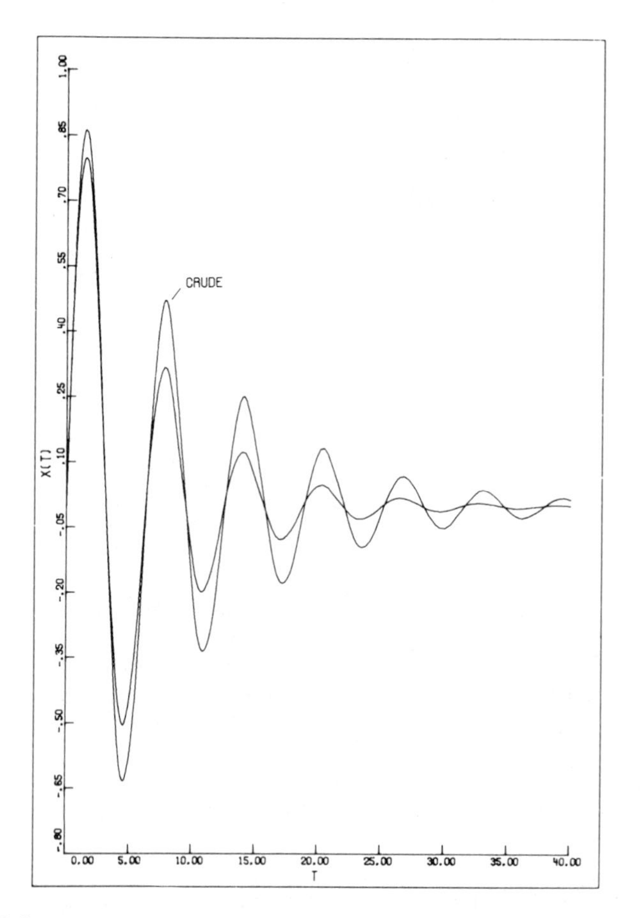


**Figure 2.7-4** *Asymptotic approximation and solution obtained by 'crude averaging' of* $\ddot{x} + 4\epsilon\cos^2(t)\dot{x} + x = 0$, $x(0)=0$, $\dot{x}(0)=1$; $\epsilon=0.1$. *The numerical solution and the asymptotic approximation nearly coincide and they decay faster than the crude 'approximation'.*

It seems perfectly natural to average the friction term to produce the equation

$$\ddot{z} + 2\epsilon\dot{z} + z = 0 \ , \ z(0)=0 \ , \ \dot{z}(0)=1.$$

We expect $z(t)$ to be an approximation of $x(t)$ on some time-scale. We have

$$z(t)=\frac{1}{(1-\epsilon^2)^{1/2}}e^{-\epsilon t}\sin((1-\epsilon^2)^{1/2}t).$$

It turns out that this is a poor result. To see this we do the averaging via the standard form as in example 2.7.1. We find

$$\frac{d\tilde{r}}{dt}=-\epsilon\tilde{r}(1-{1/2}\cos(2\tilde{\psi})\ ,\ \tilde{r}(0)=1,$$

$$\frac{d\tilde{\psi}}{dt}=-{1/2}\epsilon\sin(2\tilde{\psi})\qquad ,\ \tilde{\psi}(0)=-\frac{\pi}{2}$$

and we have $\tilde{r}(t)=e^{-\frac{3}{2}\epsilon t}$, $\tilde{\psi}(t)=-\frac{\pi}{2}$. So

$$x(t)=e^{-\frac{3}{2}\epsilon t}\sin(t)+O(\epsilon)$$

on the time-scale $\frac{1}{\epsilon}$.

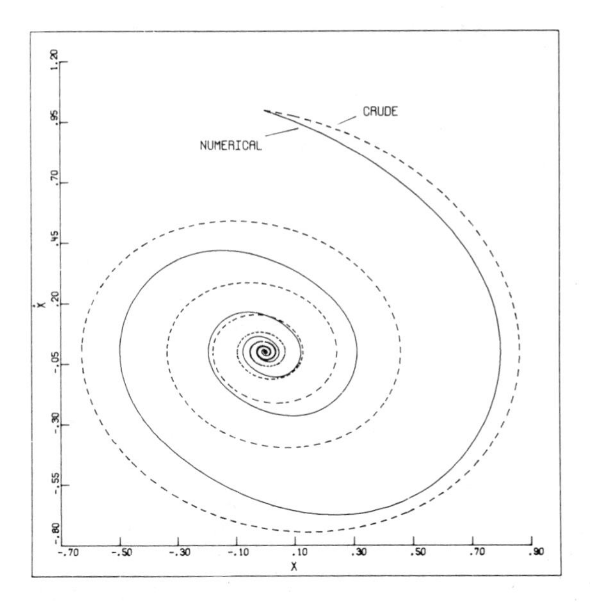


**Figure 2.7-5** *Phase-plane for* $\ddot{x}+4\epsilon\cos^2(t)\dot{x}+x=0$, $x(0)=0$, $\dot{x}(0)=1$; $\epsilon=0.1$. *The phase-orbit of the numerical solution and the asymptotic approximation nearly coincide and has been represented by a full line; the crude approximation has been indicated by* $-\ -\ -$.

Actually we shall prove in chapter 4 that this estimate is valid on $[0,\infty)$. We have clearly

$$x(t)-z(t)=O_S(1)$$

on the time-scale $\frac{1}{\epsilon}$. In figure 2.7-4 we depict $z(t)$ and $x(t)$ obtained by numerical integration. We could have plotted $e^{-\frac{3}{2}\epsilon t}\sin(t)$ but this asymptotic approximation nearly coincides with the numerical solution. It turns out that if $\epsilon=0.1$

$$\sup_{t\geqslant 0}|x(t)-e^{-\frac{3}{2}\epsilon t}\sin(t)|\leqslant 0.015$$

In figure 2.7-5 we illustrate the behavior in the phase-plane of the crude and the numerical solution.

# 3. The Theory of Averaging

## 3.1. Introduction

This chapter will be concerned with the theory of averaging for equations in the standard form

$$\dot{x} = \epsilon f(t,x) + \epsilon^2 R(t,x,\epsilon).$$

In chapter 2 we discussed how to obtain perturbation problems in the standard form and we studied a simple version of averaging. The procedure of averaging can be found already in the works of *Lagrange* and *Laplace* who provided an intuitive justification and who used the procedure to study the problem of secular perturbations in the solar system.
To many physicists and astronomers averaging seems to be such a natural procedure that they do not even bother to justify the process. However it is important to have a rigorous approximation theory, since it is precisely the fact that averaging seems so natural which obscures the pitfalls and restrictions of the method. We find for instance incorrect results based on averaging by *Jeans* (§ 268 in (Jea28a)) who studies the two-body problem with slowly varying mass; cf. the results obtained by *Verhulst*(Ver75a).

Around 1930 we see the start of precise statements and proofs in averaging theory. A historical survey of the development of the theory from the 18$^{th}$ century until 1966 can be found in appendix 8.1. After 1966 many new results in the theory of averaging have been obtained. The main trends of this research will be reflected in the subsequent chapters. We summarize part of the recent literature here.

Many results in the theory of asymptotic approximations have been obtained in the Soviet Union from 1930 onwards. Earlier work of this school has been presented in the famous book by *Bogoliubov* and *Mitropolsky* (Bog61a) and in the survey paper by *Volosov* (Vol63a). A brief glance at the main Soviet mathematical journals shows that this school is still flourishing; it is producing results on integral manifolds, equations with retarded argument, quasi- or almost- periodic equations etc. See also the survey by *Mitropolsky* (Mit73a) and the book by *Bogoliubov* , *Mitropolsky* and *Samoilenko* (Bog76a). In 1966 *Roseau* ((Ros66a), chapter 12) presented a transparent proof of the validity of averaging in the periodic case. Different proofs for both the periodic and the general case have been provided by *Besjes* (Bes69a) and *Perko* (Per69a). In the last paper moreover the relation between averaging and the multiple time-scales method has been established.

Most of the work mentioned above is concerned with approximations on the time-scale $1/\epsilon$. Extension of the time-scale of validity is possible if for instance one studies equations leading to approximations starting inside the domain of attraction of an asymptotically stable critical point. Extensions like this were studied by *Banfi* (Ban67a) and *Banfi* and *Graffi* (Ban69a). *Eckhaus* (Eck75a) gives a detailed proof and new results for systems with attraction; later the proof could be simplified considerably by *Eckhaus* using a lemma due to *Sanchez* − *Palencia*, see (Ver76a). Results on related problems have been obtained by *Kirchgraber* and *Stiefel* (Kir78a) who study periodic solutions and the part played by invariant manifolds. Systems with attraction will be studied in the next chapter. Another type of problem where extension of the time-scale is possible is provided by systems in resonance with as an important example *Hamiltonian* systems. We present a basic result on resonance problems in § 9 which is due to *van der Burgh* (Bur74a).

In the theory of averaging of periodic systems one usually obtains $O(\epsilon)$ approximations on the time-scale $1/\epsilon$. In the case of general averaging an order function $\delta(\epsilon)$ plays a part; see § 3. The order function $\delta$ is determined by the behavior of $f(t,x)$ and its average on a long time-scale. In the original theorem by *Bogoliubov* and *Mitropolsky* an $o(1)$ estimate has been given. Implicitly however, an $O(\delta^{1/2})$ estimate has been derived in the proof. Also in the proofs by *Besjes* one obtains an $O(\epsilon^{1/2})$ estimate in the general case but, using a different proof, an $O(\epsilon)$ estimate in the periodic case.

A curiosity of the proofs is that in restricting the proof of the general case to the case of periodic systems one still obtains an $O(\epsilon^{1/2})$ estimate. It takes a special effort to find an $O(\epsilon)$ estimate for periodic systems. *Eckhaus* (Eck75a) introduces the concept of *local average* of a vectorfield to give a new proof of the validity of periodic and general averaging. In the general case *Eckhaus* finds an $O(\delta^{1/2})$ estimate; on specializing to the periodic case one can apply the averaging repeatedly to obtain an $O(\epsilon^r)$ estimate where $r$ approaches 1 from below.

In the sequel we shall use *Eckhaus*' concept of local average to derive in a simple way an $O(\epsilon)$ estimate in the periodic case under rather weak assumptions (§ 2). In the general case one finds under similar assumptions an $O(\delta^{1/2})$ estimate (§ 3).

In § 4 we present the theory of second order approximation in the general case; we find here that the first order approximation is valid with $O(\delta)$-accuracy in the general case if we require the vectorfield to be differentiable instead of only *Lipschitz*-continuous.

## 3.2. Basic lemmas; the periodic case

In this section we shall derive some basic results which are preliminary for our treatment of the general theory of averaging.

### 3.2.1. Definition (*Eckhaus*(Eck75a))

Consider the continuous vectorfield $f:\mathbf{R}\times\mathbf{R}^p\rightarrow\mathbf{R}^n$. We define the *local average* $f_T$ of $f$ by

$$f_T(t,x)=\frac{1}{T}\int_0^T f(t+\tau,x)d\tau,$$

in which $x$ is a dummy variable; $p$ is zero or a natural number, e.g. $n$.

### 3.2.2. Remark

$T$ is a parameter which can be chosen. Note that the local average of a continuous vectorfield always exists. If $f$ is T-periodic in $t$, then the local average $f_T$ equals the usual average $f^o$.

### 3.2.3. Lemma

Consider the continuous vectorfield $f:\mathbf{R}\times\mathbf{R}^p\rightarrow\mathbf{R}^n$, T-periodic in $t$. Then

$$f_T(t,x)=f^o(x)\ (\ =\ \frac{1}{T}\int_0^T f(t,x)dt).$$

**Proof**

We write $f_T(t,x)=\frac{1}{T}\int_t^{t+T} f(\tau,x)d\tau$. Partial differentiation with respect to $t$ produces zero because of the $T$-periodicity of $f$; $f_T$ does not depend *explicitly* on $t$, so we may put $t=0$. □

We shall now introduce vectorfields which can be averaged in a general sense. As most applications are for differential equations we impose some additional regularity conditions on the vectorfield.

**3.2.4. Definition**

Consider the vectorfield $f(t,x)$ with $f:\mathbf{R}\times\mathbf{R}^n\to\mathbf{R}^n$, *Lipschitz*-continuous in $x$ on $D\subset\mathbf{R}^n$, $t\geqslant 0$; $f$ continuous in $t$ and $x$ on $\mathbf{R}^+\times D$. If the average

$$f^o(x)=\lim_{T\to\infty}\frac{1}{T}\int_0^T f(t,x)dt$$

exists, $f$ is called a *KBM-vectorfield* ( *KBM stands for Krylov, Bogoliubov* and *Mitropolsky*).

**3.2.5. Remark**

The part played by $T$ is a special one. Whenever we estimate with respect to $\epsilon$ we shall require $\epsilon T=o(1)$, where $\epsilon$ is a small parameter. So we may choose for instance $T=\frac{1}{\epsilon^{1/2}}$ or $T=\frac{1}{\epsilon|\log(\epsilon)|}$. Included is also $T=O_s(1)$ which we shall use for periodic $f$. We now have a simple estimate by

**3.2.6. Lemma**

Consider the *Lipschitz*-continuous map $\phi:\mathbf{R}\to\mathbf{R}^n$, then

$$\phi(t)=\phi_T(t)+O(T).$$

**3.2.7. Remark**

Here and in the sequel the interpretation of the estimate $O(T)$ is such that the estimate holds for the argument of the symbol $O(.)$ tending to any value in $\mathbf{R}\cup\{\pm\infty\}$.

**Proof**

Let $\lambda$ be the *Lipschitz*-constant of $\phi$. Then

$$|\phi(t)-\phi_T(t)|=|\frac{1}{T}\int_0^T(\phi(t)-\phi(t+\tau))d\tau|\leqslant\frac{1}{T}\int_0^T\lambda\tau d\tau=\tfrac{1}{2}\lambda T.\quad\square$$

In the following lemma we introduce a perturbation problem in the standard form.

**3.2.8. Lemma**

Consider the equation

$$\dot{x}=\epsilon f(t,x)\quad t\geqslant 0\ ,\ x\in D\subset\mathbf{R}^n.$$

Assume

$$|f(t,x)-f(t,y)|\leqslant\lambda|x-y|$$

for all $x,y\in D$ ( *Lipschitz*-continuity ) and $f$ is continuous in $t$ and $x$; Let

$$M = \sup_{x \in D} \sup_{0 \leqslant \epsilon t \leqslant L} |f(t,x)| < \infty.$$

The constants $\lambda$, $L$ and $M$ are $\epsilon$-independent. If

$$\phi(t) = \int_0^t f(\tau, x(\tau)) d\tau,$$

where $x$ is a solution of the differential equation, then we have, with $t$ on the time-scale $\frac{1}{\epsilon}$

$$|\phi_T - \int_0^t f_T(\tau, x(\tau)) d\tau| \leqslant \tfrac{1}{2}(1 + \lambda L) MT$$

or

$$\phi_T(t) = \int_0^t f_T(\tau, x(\tau)) d\tau + O(T).$$

**Proof**

By definition

$$\phi_T(t) = \frac{1}{T} \int_0^T \int_0^{t+\tau} f(\sigma, x(\sigma)) d\sigma d\tau = \frac{1}{T} \int_0^T \int_\tau^{t+\tau} f(\sigma, x(\sigma)) d\sigma d\tau + R_1$$

$$= \frac{1}{T} \int_0^T \int_0^t f(\sigma + \tau, x(\sigma + \tau)) d\sigma d\tau + R_1 = \frac{1}{T} \int_0^t \int_0^T f(\sigma + \tau, x(\sigma)) d\tau d\sigma + R_1 + R_2$$

$$= \int_0^t f_T(\tau, x(\tau)) d\tau + R_1 + R_2.$$

$R_1$ and $R_2$ have been defined implicitly and we estimate these quantities as follows:

$$|R_1| = |\frac{1}{T} \int_0^T \int_0^\tau f(\sigma, x(\sigma)) d\sigma d\tau| \leqslant \frac{1}{T} \int_0^T \int_0^\tau M d\sigma d\tau \leqslant \tfrac{1}{2} MT$$

and

$$|R_2| = |\frac{1}{T} \int_0^t \int_0^T [f(\sigma + \tau, x(\sigma + \tau)) - f(\sigma + \tau, x(\sigma))] d\tau d\sigma|$$

$$\leqslant \frac{\lambda}{T} \int_0^t \int_0^T |x(\sigma + \tau) - x(\sigma)| d\tau d\sigma$$

$$\leqslant \epsilon \frac{\lambda}{T} \int_0^t \int_0^T \int_\sigma^{\sigma + \tau} |f(\sigma', x(\sigma'))| d\sigma' d\tau d\sigma$$

$$\leqslant \epsilon \frac{\lambda}{T} \int_0^t \int_0^T M \tau d\tau d\sigma = \tfrac{1}{2} \epsilon t \lambda MT = O(T)$$

for $t$ on the time-scale $\frac{1}{\epsilon}$.□

The preceding lemmas enable us to compare solutions of two differential equations:

**3.2.9. Lemma**

Consider the initial value problem

$$\dot{x}=\epsilon f(t,x)\ ,\ x(0)=x_o,$$

with $f:\mathbf{R}\times\mathbf{R}^n$ *Lipschitz*-continuous in $x$ on $D\subset\mathbf{R}^n$, $t$ on the time-scale $\frac{1}{\epsilon}$; $f$ continuous in $t$ and $x$. If $y$ is the solution of

$$\dot{y}=\epsilon f_T(t,y)\ ,\ y(0)=x_o,$$

then $x(t)=y(t)+O(\epsilon T)$ on the time-scale $\frac{1}{\epsilon}$.

**Proof**

$$x(t)=x_o+\epsilon\int_0^t f(\tau,x(\tau))d\tau.$$

Now with Lemma 3.2.6

$$\phi(t)=\int_0^t f(\tau,x(\tau))d\tau=\phi_T(t)+O(T)$$

(and with Lemma 3.2.8)

$$=\int_0^t f_T(\tau,x(\tau))d\tau+O(T).$$

So $x(t)=x_o+\epsilon\int_0^t f_T(\tau,x(\tau))d\tau+O(\epsilon T)$. Since

$$y(t)=x_o+\epsilon\int_0^t f_T(\tau,y(\tau))d\tau$$

we have

$$x(t)-y(t)=\epsilon\int_o^t[f_T(\tau,x(\tau))-f_T(\tau,y(\tau))]d\tau+O(\epsilon T)$$

and because of the *Lipschitz*-continuity of $f_T$ (inherited from $f$)

$$|x(t)-y(t)|\leqslant\epsilon\int_o^t\lambda|x(\tau)-y(\tau)|d\tau+O(\epsilon T).$$

*Gronwall's* lemma yields

$$|x(t)-y(t)|=O(\epsilon Te^{\epsilon\lambda t}),$$

from which the lemma follows.□

At this stage it is a trivial application of Lemma 3.2.3 and Lemma 3.2.9 to prove the averaging theorem in the periodic case (Cf. Theorem 2.6.1). We formulate the theorem, based on this new proof; note that we used assumption (b) implicitly in the proof of Lemma 3.2.9.

**3.2.10. Theorem (periodic averaging)**

Consider the initial value problems

$$\dot{x}=\epsilon f(t,x)\ ,\ x(0)=x_o,$$

with $f{:}\mathbf{R}^{n+1}\rightarrow\mathbf{R}^n$ and

$$\dot{y}=\epsilon f^o(y)\ ,\ y(0)=x_o,$$

$x,y,x_o\in D\subset\mathbf{R}^n$, $t\in[0,\infty)$, $\epsilon\in(0,\epsilon_o]$. Suppose

a) $f$ has period $T$;

b) $f$ is *Lipschitz*-continuous in $x$ on $D\subset\mathbf{R}^n$, $t\geqslant 0$, continuous in $t$ and $x$ on $\mathbf{R}^+\times D$ and with average $f^o$;

c) $y(t)$ belongs to an interior subset of $D$ on the time-scale $\frac{1}{\epsilon}$;

then

$$x(t)-y(t)=O(\epsilon)\ \textit{as}\ \epsilon\downarrow 0\ \textit{on the time-scale}\ \frac{1}{\epsilon}.$$

**Remark**

From assumptions (a) and (b) it follows that $f$ is a *KBM*-vectorfield. The only gain until now is that we weakened the assumption of differentiability in Theorem 2.6.1 to *Lipschitz*-continuity in the periodic case by using Lemma 3.2.9. Though the proof looks different it has in fact been based on *Gronwall*'s lemma 1.3.1 as in the proof in chapter 2. Serious progress however can be made in the general case where we shall obtain sharp estimates while keeping the same kind of simple proofs as in this section.

## 3.3. General averaging

To prove the fundamental theorem of general averaging we need a few more results.

**3.3.1. Lemma**

If $f$ is a *KBM*-vectorfield then

$$f_T(t,x)=f^o(x)+O(\frac{\delta(\epsilon)}{\epsilon T}),$$

where

$$\delta(\epsilon)=\sup_{x\in D}\ \sup_{t\in[0,\frac{L}{\epsilon})}\epsilon\left|\int_0^t[f(\tau,x)-f^o(x)]d\tau\right|$$

and

$$\epsilon T = o(1) \ as \ \epsilon \to 0.$$

**Remark**

We remind the reader of the interpretation of the $O$-symbol in this context; see Remark 3.2.7.

**Proof**

$$f_T(t,x) - f^o(x) = \frac{1}{T}\int_0^T [f(t+\tau,x) - f^o(x)]d\tau$$

$$= \frac{1}{T}\int_0^{t+T} [f(\tau,x) - f^o(x)]d\tau - \frac{1}{T}\int_0^t [f(\tau,x) - f^o(x)]dtau.$$

We assumed $\epsilon T = o(1)$, so if $\alpha = 0$ or $T$ we have $\epsilon\alpha = o(1)$ and

$$\left| \int_0^{t+\alpha} [f(\tau,x) - f^o(x)]d\tau \right| \leqslant \frac{\delta(\epsilon)}{\epsilon},$$

from which the estimate follows. □

**3.3.2. Lemma**

Let $y$ be the solution of the initial value problem

$$\dot{y} = \epsilon f_T(t,y) \ , \ y(0) = x_o.$$

We suppose $f$ is a *KBM*-vectorfield; $z$ is the solution of the initial value problem

$$\dot{z} = \epsilon f^o(z) \ , \ z(0) = x_o;$$

then

$$y(t) = z(t) + O(\frac{\delta(\epsilon)}{\epsilon T}),$$

with $t$ on the time-scale $\frac{1}{\epsilon}$.

**Proof**

$$y(t) - z(t) = \epsilon\int_0^t [f_T(\tau, y(\tau)) - f^o(z(\tau))]d\tau$$

(and with Lemma 3.3.1)

$$= \epsilon\int_0^t [f^o(y(\tau)) - f^o(z(\tau))]d\tau + O(\frac{\delta(\epsilon)}{T}t).$$

Using the *Lipschitz*-constant $\lambda$ of $f^o$ we obtain

$$|y(t)-z(t)| \leqslant O(\frac{\delta(\epsilon)}{\epsilon T} e^{\epsilon \lambda t})$$

from which the Lemma follows. □

We are now able to prove the general averaging theorem:

**3.3.3. Theorem (general averaging)**

Consider the initial value problems

$$\dot{x} = \epsilon f(t,x) \ , \ x(0) = xo,$$

with $f{:}\mathbf{R}^{n+1} \to \mathbf{R}^n$ and

$$\dot{y} = \epsilon f^o(y) \ , \ y(0) = x_o,$$

$x,y,x_o \in D \subset \mathbf{R}^n$ , $t \in [0,\infty)$, $\epsilon \in (0,\epsilon_o]$. Suppose

a) f is a *KBM*-vectorfield with average $f^o$;

b) $y(t)$ belongs to an interior subset of $D$ on the time-scale $\frac{1}{\epsilon}$;

then

$$x(t) - y(t) = O(\delta^{\frac{1}{2}}(\epsilon)) \text{ as } \epsilon \to 0 \text{ on the time-scale } \frac{1}{\epsilon},$$

where

$$f^o(x) = \lim_{T \to \infty} \frac{1}{T} \int_0^T f(t,x) dt,$$

$$\delta(\epsilon) = \sup_{x \in D} \ \sup_{t \in [0,\frac{L}{\epsilon})} \epsilon \left| \int_0^t [f(\tau,x) - f^o(x)] d\tau \right| .$$

**Proof**

Applying Lemma 3.2.9 and 3.3.2, using the triangle inequality, we have on the time-scale $\frac{1}{\epsilon}$:

$$x(t) = y(t) + O(\epsilon T) + O(\frac{\delta(\epsilon)}{\epsilon T}).$$

The errors are of the same order of magnitude if

$$\epsilon^2 T^2 = \delta(\epsilon),$$

so that

$$x(t) = y(t) + O(\delta^{\frac{1}{2}}(\epsilon)).$$

□

**3.3.4. Remark**

As before, condition (b) has been used implicitly in the estimates. Note that if $\delta(\epsilon)=o(1)$ the $\epsilon T=o(1)$; of course if $\delta(\epsilon)\neq o(1)$ the estimates are useless.

To understand the theory of general averaging it is instructive to analyze a few simple examples.

**3.3.5. Example: linear oscillator with increasing damping**

Consider the equation

$$\ddot{x}+\epsilon(2-F(t))\dot{x}+x=0,$$

with initial values given at $t=0$ : $x(0)=r_o$, $\dot{x}(0)=0$. $F(t)$ is a continuous function, monotonically decreasing towards zero for $t\rightarrow\infty$ with $F(0)=1$. So the problem is simple: we start with an oscillator with damping coefficient $\epsilon$, we end up (in the limit for $t\rightarrow\infty$) with an oscillator with damping coefficient $2\epsilon$. We shall show that on the time-scale $\frac{1}{\epsilon}$ the system behaves approximately as if it has the limiting damping coefficient $2\epsilon$, which seems an interesting result. To obtain the standard form, transform $(x,\dot{x})\mapsto(r,\phi)$ by

$$x=r\cos(t+\phi)\ ,\ \dot{x}=-r\sin(t+\phi).$$

We find

$$\dot{r}=\epsilon r\sin^2(t+\phi)(-2+F(t)) \qquad ,\ r(0)=r_o,$$

$$\dot{\phi}=\epsilon\sin(t+\phi)\cos(t+\phi)(-2+F(t))\ ,\ \phi(0)=0.$$

Averaging produces

$$\dot{\tilde{r}}=-\epsilon\tilde{r}\ ,\ \dot{\tilde{\phi}}=0,$$

so that

$$x(t)=r_o e^{-\epsilon t}\cos(t)+O(\delta^{1/2}(\epsilon)),$$

$$\dot{x}(t)=-r_o e^{-\epsilon t}\sin(t)+O(\delta^{1/2}(\epsilon)).$$

To estimate $\delta$ we note that $x=\dot{x}=0$ is a globally stable attractor (one can use the *Lyapunov*-function $\frac{1}{2}(x^2+\dot{x}^2)$ to show this if the mechanics of the problem is not already convincing enough). So the order of magnitude of $\delta$ is determined by

$$\sup_{t\in[0,\frac{L}{\epsilon})}\epsilon\left|\int_0^t[\sin^2(\tau+\phi)(-2+F(\tau))+1]d\tau\right|$$

and

$$\sup_{t\in[0,\frac{L}{\epsilon})}\epsilon\left|\int_0^t[\sin(\tau+\phi)\cos(\tau+\phi)(-2+F(\tau))]d\tau\right|.$$

The second integral is bounded for all $t$ so this contributes $O(\epsilon)$. The same holds for the part

$$\int_o^t (-2\sin^2(\tau+\phi)+1)d\tau.$$

To estimate

$$\int_0^t F(\tau)\sin^2(\tau+\phi)d\tau$$

we have to make an assumption about $F$. For instance if $F$ decreases exponentially with time we have $\delta(\epsilon)=O(\epsilon)$ and an approximation with error $O(\epsilon^{1/2})$. If $F \sim t^{-s}(0<s<1)$ we have $\delta(\epsilon)=O(\epsilon^s)$ and an approximation with error $O(\epsilon^{\frac{s}{2}})$. If $F(t)=(1+t)^{-1}$ we have $\delta(\epsilon)=O(\epsilon\,|\log(\epsilon)|)$. We remark finally that to describe the dependence of the oscillator on the initial damping we clearly need a different order of approximation.

In discussing higher order approximations we shall prove that the error of the first order approximation is $O(\delta)$ instead of $O(\delta^{1/2}(\epsilon))$ ( see § 5 ).

## 3.4. Second order approximations in general averaging; an improved first-order estimate assuming differentiability

Higher order approximations in the periodic case are well known and form an established theory with many applications. We are not aware of similar results in the general case and we shall discuss the subject in this section. It turns out there is an unexpected profit: the first order approximation is better than we proved it to be in § 3 under the differentiability condition.

### 3.4.1. Lemma

Suppose $f$ is a *KBM*-vectorfield which has a *Lipschitz*-continuous first derivative in $x$; $x \in D \subset \mathbf{R}^n$, $t$ on the time-scale $\frac{1}{\epsilon}$; $x$ is the solution of

$$\dot{x}=\epsilon f(t,x)\ ,\ x(0)=x_o.$$

We define $w$ by

$$x(t)=w(t)+\delta_1(\epsilon)u^1(t,w(t)),$$

where

$$\delta_1(\epsilon)u^1(t,w)=\epsilon\int_0^t [f(\tau,w)-f^o(w)]d\tau.$$

Then

$$w(t)=x_o+\epsilon\int_0^t f^o(w(\tau))d\tau+\epsilon\delta_1(\epsilon)\int_0^t [\nabla f(\tau,w(\tau))u^1(\tau,w(\tau))$$

$$-\nabla u^1(\tau,w(\tau))f^o(w(\tau))]d\tau+O(\delta_1^2)$$

on the time-scale $\frac{1}{\epsilon}$ ( as before $\nabla g(t,w)$ indicates the first derivative of the vectorfield $g$ with respect to $w$ ).

**Proof**

This is a standard computation:

$$w(t)=x(t)-\delta_1(\epsilon)u^1(t,w(t))$$

$$=x_o+\epsilon\int_0^t f(\tau,x(\tau))d\tau-\epsilon\int_0^t[f(\tau,w(\tau))-f^o(w(\tau))]d\tau$$

$$-\delta_1(\epsilon)\int_0^t\nabla u^1(\tau,w(\tau))\frac{dw}{dt}d\tau=x_o+\epsilon\int_0^t f^o(w(\tau))d\tau$$

$$+\epsilon\delta_1(\epsilon)\int_0^t[\nabla f(\tau,w(\tau))u^1(\tau,w(\tau))-\nabla u^1(\tau,w(\tau))f^o(w(\tau))]d\tau+O(\delta_1^2)$$

□

**3.4.2. Lemma**

Let $w$ be defined as in Lemma 3.4.1 and let $v$ be the solution of

$$\dot v=\epsilon f^o(v)+\epsilon\delta_1(\epsilon)f_T^1(t,v)\ ,\ v(0)=x_o,$$

where

$$f^1(t,v)=\nabla f(t,v)u^1(t,v)-\nabla u^1(t,v)f^o(v).$$

*Assume that $f^1$ is a KBM-vectorfield,* then

$$w(t)=v(t)+O(\delta_1(\epsilon)[\epsilon T+\delta_1(\epsilon)])$$

on the time-scale $\frac{1}{\epsilon}$.

**Proof**

$$v(t)=x_o+\epsilon\int_0^t f^o(v(\tau))d\tau+\epsilon\delta_1(\epsilon)\int_0^t f_T^1(\tau,v(\tau))d\tau.$$

In the same way as in the proof of Lemma 3.2.8 we find

$$\int_0^t f_T^1(\tau,v(\tau))d\tau=\int_0^t f^1(\tau,v(\tau))d\tau+O(T),$$

so

$$v(t)=x_o+\epsilon\int_0^t f^o(v(\tau))d\tau+\epsilon\delta_1(\epsilon)\int_0^t f^1(\tau,v(\tau))d\tau+O(\delta_1(\epsilon)T).$$

Subtracting this from the estimate for $w$ in Lemma 3.4.1 produces

$$w(t)-v(t)=\epsilon\int_0^t[f^o(w(\tau))-f^o(v(\tau))]d\tau$$

$$+\epsilon\delta_1(\epsilon)\int_0^t[f^1(\tau,w(\tau))-f^1(\tau,v(\tau))]d\tau+O(\delta_1^2(\epsilon)+\delta_1\epsilon T).$$

Using the *Lipschitz*-continuity of $f^o$ and $f^1$ and applying the *Gronwall*-lemma yields the desired result. □

For the analysis of second order approximations we need one more lemma

**3.4.3. Lemma**

Let $u$ (not to be mistaken for $u^1$) be the solution of

$$\dot{u}=\epsilon f^o(u)+\epsilon\delta_1(\epsilon)f^{1o}(u)\ ,\ u(0)=x_o$$

($f^{1o}$ is the general average of $f^1$, which is assumed to be *KBM*) and

$$\delta_2(\epsilon)=\sup_{x\in D}\ \sup_{t\in[0,\frac{L}{\epsilon})}\epsilon\left|\int_0^t[f^1(\tau,x)-f^{1o}(x)]d\tau\right|.$$

Let $v$ be defined as in Lemma 3.4.2, then

$$v(t)=u(t)+O(\frac{\delta_1(\epsilon)\delta_2(\epsilon)}{\epsilon T})$$

on the time-scale $\frac{1}{\epsilon}$.

**Proof**

$$v(t)-u(t)=\epsilon\int_0^t[f^o(v(\tau))-f^o(u(\tau))]d\tau+\epsilon\delta_1(\epsilon)\int_0^t[f_T^1(\tau,v(\tau))-f^{1o}(u(\tau))]d\tau.$$

With Lemma 3.3.1

$$f_T^1(t,v(t))=f^{1o}(v(t))+O(\frac{\delta_2(\epsilon)}{\epsilon T})$$

and the *Lipschitz*-continuity of $f^o$ and $f^{1o}$ ( *Lipschitz*-constants $\lambda_1$ and $\lambda_2$ )

$$|v(t)-u(t)|\leqslant\epsilon\lambda_1\int_0^t|v(\tau)-u(\tau)|d\tau+\epsilon\delta_1(\epsilon)\lambda_2\int_0^t|v(\tau)-u(\tau)|d\tau$$

$$+O(\frac{\delta_1(\epsilon)\delta_2(\epsilon)}{\epsilon T}\epsilon t).$$

Application of the *Gronwall*-lemma produces the estimate of the Lemma. □

**3.4.4. Theorem (second order approximation in general averaging)**

Consider the initial value problems

$$\dot{x}=\epsilon f(t,x)\ ,\ x(0)=x_o$$

and

$$\dot{u}=\epsilon f^o(u)+\epsilon\delta_1(\epsilon)f^{1o}(u)\ ,\ u(0)=x_o,$$

with $f:\mathbf{R}\times\mathbf{R}^n\rightarrow\mathbf{R}^n$; $x,u,x_o\in D\subset\mathbf{R}_n$, $t\in[0,\infty)$, $\epsilon\in(0,\epsilon_o]$, and

$$f^1(t,x)=\nabla f(t,x)u^1(t,x)-\nabla u^1(t,x)f^o(x),$$

where

$$\delta_1(\epsilon)u^1(t,x)=\epsilon\int_0^t[f(\tau,x)-f^o(x)]d\tau$$

and

$$\delta_1(\epsilon)=\sup_{x\in D}\ \sup_{t\in[0,\frac{L}{\epsilon})}\epsilon\left|\int_0^t[f(\tau,x)-f^o(x)]d\tau\right|.$$

Suppose

a) $f$ and $f^1$ are *KBM*-vectorfields (with average $f^o$ and $f^{1o}$);

b) $u(t)$ belongs to an interior subset of $D$ on the time-scale $\frac{1}{\epsilon}$;

then on the time-scale $\frac{1}{\epsilon}$

$$x(t)=u(t)+\delta_1(\epsilon)u^1(t,u(t))+O(\delta_1(\epsilon)[\delta_2^{1/2}(\epsilon)+\delta_1(\epsilon)]),$$

with

$$\delta_2(\epsilon)=\sup_{x\in D}\ \sup_{t\in[0,\frac{L}{\epsilon})}\epsilon\left|\int_0^t[f^1(\tau,x)-f^{1o}(x)]d\tau\right|.$$

**Proof**

With $w$ defined as in Lemma 3.4.1 we have

$$\begin{aligned}&|x(t)-(u(t)+\delta_1(\epsilon)u^1(t,u(t)))|\\&=|w(t)+\delta_1(\epsilon)u^1(t,w(t))-(u(t)+\delta_1(\epsilon)u^1(t,u(t)))|\\&\leqslant(1+\delta_1(\epsilon))\,|w(t)-u(t)|,\end{aligned}$$

where we used the triangle inequality and the *Lipschitz*-continuity of $u^1$(constant $\lambda$). Again using the triangle inequality and Lemma 3.4.2 and 3.4.3 we find

$$|w(t)-u(t)|=O(\delta_1(\epsilon)[\epsilon T+\delta_1(\epsilon)])+O(\frac{\delta_1(\epsilon)\delta_2(\epsilon)}{\epsilon T}).$$

We choose $T$ such that the errors are of the same order, so $\epsilon^2 T^2 = \delta_2(\epsilon)$. This choice produces the estimate of the theorem.□

A remarkable consequence of this theorem is an improved estimate for the first-order result of Theorem 3.3.3. However, this is an improvement obtained after making additional assumptions.

**3.4.5. Theorem**

Consider the initial value problems

$$\dot{x} = \epsilon f(t,x)\ ,\ x(0) = x_o$$

and

$$\dot{y} = \epsilon f^o(y)\ ,\ y(0) = x_o;$$

then, with the assumptions of Theorem 3.4.4,

$$x(t) = y(t) + O(\delta_1(\epsilon)),$$

where

$$\delta_1(\epsilon) = \sup_{x \in D} \sup_{t \in [0, \frac{L}{\epsilon})} \epsilon \left| \int_0^t [f(\tau,x) - f^o(x)] d\tau \right| .$$

**Proof**

With the *Gronwall*-lemma we have in the usual way

$$u(t) = y(t) + O(\delta_1(\epsilon))$$

on the time-scale $\frac{1}{\epsilon}$; $u$ has been defined in Theorem 3.4.4. Also from Theorem 3.4.4 we have

$$x(t) = u(t) + O(\delta_1(\epsilon)).$$

The triangle inequality produces the desired result.□

An extension of Theorem 3.4.4 which is nearly trivial but useful can be made as follows.

**3.4.6. Theorem**

Consider the initial value problems

$$\dot{x} = \epsilon f(t,x) + \epsilon^2 g(t,x) + \epsilon^3 R(t,x,\epsilon)\ ,\ x(0) = x_o$$

and

$$\dot{u} = \epsilon f^o(u) + \epsilon\delta_1(\epsilon) f^{1o}(u) + \epsilon^2 g^o(u)\ ,\ u(0) = x_o,$$

with $f,g: \mathbf{R} \times \mathbf{R}^n \to \mathbf{R}^n$, $R: \mathbf{R} \times \mathbf{R}^n \times (0,\epsilon_o] \to \mathbf{R}^n$, $x,u,x_o \in D \subset \mathbf{R}^n$, $t \in [0,\infty)$ and $\epsilon \in (0,\epsilon_o]$ with

$$f^1(t,x) = \nabla f(t,x) u^1(t,x) - \nabla u^1(t,x) f^o(x),$$

where

$$\delta_1(\epsilon)u^1(t,w)=\epsilon\int_0^t[f(\tau,w)-f^o(w)]d\tau$$

and

$$\delta_1(\epsilon)=\sup_{x\in D}\ \sup_{t\in[0,\frac{L}{\epsilon})}\epsilon\left|\int_0^t[f(\tau,x)-f^o(x)]d\tau\right|.$$

Suppose

a) $f,f^1$ and $g$ are *KBM*-vectorfields, with averages $f^o,f^{1o}$ and $g^o$; $|R(t,x,\epsilon)|$ is bounded by a constant uniformly on $[0,\frac{L}{\epsilon})\times D\times(0,\epsilon_o]$;

b) $u(t)$ belongs to an interior subset of $D$ on the time-scale $\frac{1}{\epsilon}$;

then on the time-scale $\frac{1}{\epsilon}$

$$x(t)=u(t)+\delta_1(\epsilon)u^1(t,u(t))+O(\delta_1(\epsilon)[\delta_2^{1/2}(\epsilon)+\delta_1(\epsilon)]+\delta_3^{1/2}(\epsilon)),$$

with

$$\delta_2(\epsilon)=\sup_{x\in D}\ \sup_{t\in[0,\frac{L}{\epsilon})}\epsilon\left|\int_0^t[f^1(\tau,x)-f^{1o}(x)]d\tau\right|,$$

$$\delta_3(\epsilon)=\sup_{x\in D}\ \sup_{t\in[0,\frac{L}{\epsilon})}\epsilon^2\left|\int_0^t[g(\tau,x)-g^o(x)]d\tau\right|.$$

**Proof**

With some small additions, the proof of Theorem 3.4.4 can be repeated. □

### 3.4.7. Example of second order averaging

We return to the linear oscillator with increasing damping (Example 3.3.5):

$$\ddot{x}+\epsilon(2-F(t))\dot{x}+x=0 \qquad , x(0)=r_o\ , \ \dot{x}(0)=0,$$

$F(0)=1$ and $F(t)$ decreases monotonically towards zero. Transforming $x,\dot{x}\mapsto r,\phi$ gives

$$\dot{r}=\epsilon r\sin^2(t+\phi)(-2+F(t)) \qquad , r(0)=r_o,$$

$$\dot{\phi}=\epsilon\sin(t+\phi)\cos(t+\phi)(-2+F(t))\ ,\ \phi(0)=0.$$

General averaging produced the vectorfield $f^o(r,\phi)=(-r,0)$. First we suppose

$$f(t)=\frac{1}{(1+t)^s}\ ,\ s>0.$$

We have

$$\delta_1(\epsilon)=O(\epsilon) \quad , s>1,$$
$$\delta_1(\epsilon)=O(\epsilon\log(\epsilon)) \quad , s=1,$$
$$\delta_1(\epsilon)=O(\epsilon^s) \quad , 0<s<1.$$

Now in the notation of Theorem 3.4.4

$$\nabla f=\begin{bmatrix} \frac{1}{2}[1-\cos(2(t+\phi))][-2+F(t)] & r\sin(2(t+\phi))[-2+F(t)] \\ 0 & \cos(2(t+\phi))[-2+F(t)] \end{bmatrix},$$

$$\delta_1(\epsilon)u^1=\epsilon\begin{bmatrix} \frac{r}{2}\int_0^t [F(\tau)-F(\tau)\cos(2(\tau+\phi))+2\cos(2(\tau+\phi))]d\tau \\ \int_0^t [-\sin(2(\tau+\phi))+\frac{1}{2}F(\tau)\sin(2(\tau+\phi))]d\tau \end{bmatrix}$$

$$=\epsilon\begin{bmatrix} \frac{r}{2}I_1(t,\phi) \\ I_2(t,\phi) \end{bmatrix} (\textit{ which defines } I_1 \textit{ and } I_2),$$

$$\delta_1(\epsilon)\nabla u^1=\epsilon\begin{bmatrix} \frac{1}{2}I_1(t,\phi) & \frac{r}{2}\frac{\partial}{\partial\phi}I_1(t,\phi) \\ 0 & \frac{\partial}{\partial\phi}I_2(t,\phi) \end{bmatrix}.$$

To compute $f^{1o}$ we have to average $\nabla f u^1$ and $\nabla u^1 f^o$. It is easy to see that for $f^{1o}$ to exist we have the condition $s>1$. So if $f(t)$ does not decrease fast enough ($0<s\leqslant 1$) the second order approximation in the sense of Theorem 3.4.4 does not exist. The calculation of the second order approximation in the case $s>1$ involves long expressions which we omit. We finally discuss the case $F(t)=e^{-t}$; note that $\delta_1(\epsilon)=O(\epsilon)$. Again in the notation of Theorem 3.4.4 we have the same expressions as before, except that now

$$I_1(t,\phi)=\sin(2(t+\phi))-e^{-t}+1+\frac{1}{5}e^{-t}\cos(2(t+\phi))$$
$$-\frac{1}{5}\cos(2\phi)-\frac{2}{5}e^{-t}\sin(2(t+\phi))-\frac{3}{5}\sin(2\phi)$$

and

$$I_2(t,\phi)=\frac{1}{2}\cos(2(t+\phi))-\frac{1}{10}e^{-t}\sin(2(t+\phi))$$
$$-\frac{1}{5}e^{-t}\cos(2(t+\phi))+\frac{1}{10}\sin(2\phi)-\frac{3}{10}\cos(2\phi).$$

After calculating $f^1$ and averaging we find simply

$$f^{1o}=(0,-\frac{1}{2}),$$

so if $u$ in Theorem 3.4.4 is written $(\overline{r},\overline{\phi})$

$$\dot{\bar{r}} = -\epsilon\bar{r}\ ,\ \bar{r}(0) = r_o,$$

$$\dot{\bar{\phi}} = -\tfrac{1}{2}\epsilon^2\ ,\ \bar{\phi}(0) = 0$$

and $\bar{r}(t) = r_o e^{-\epsilon t}$, $\bar{\phi}(t) = -\tfrac{1}{2}\epsilon^2 t$.

For the solution of the original perturbation problem we have

$$x(t) =$$

$$[r_o e^{-\epsilon t} + \frac{\epsilon}{2} r_o e^{-\epsilon t} I_1(t, -\tfrac{1}{2}\epsilon^2 t)]\cos(t - \tfrac{1}{2}\epsilon^2 t + \epsilon I_2(t, -\tfrac{1}{2}\epsilon^2 t)) + O(\epsilon^{\frac{3}{2}})$$

on the time-scale $\frac{1}{\epsilon}$.

## 3.5. Second order approximation in the periodic case

Applying the general averaging Theorem 3.3.3 to the periodic case produces an estimate of $O(\epsilon^{\frac{1}{2}})$ on the time-scale $\frac{1}{\epsilon}$ instead of the $O(\epsilon)$ estimate of Theorem 3.2 10. The same type of phenomenon occurs in second order approximations. In the periodic case we have in Theorem 3.4.4 $\delta_1(\epsilon) = \delta_2(\epsilon) = \epsilon$ and the estimate becomes of order $\epsilon^{\frac{3}{2}}$ on the time-scale $\frac{1}{\epsilon}$. Again, improvement of the estimate is possible without additional assumptions.

### 3.5.1. Theorem (second order approximation in the periodic case)

Consider the initial value problems

$$\dot{x} = \epsilon f(t,x) + \epsilon^2 g(t,x) + \epsilon^3 R(t,x,\epsilon)\ ,\ x(0) = x_o$$

and

$$\dot{u} = \epsilon f^o(u) + \epsilon^2 f^{1o}(u) + \epsilon^2 g^o(u)\ ,\ u(0) = x_o,$$

with $f,g:[0,\infty)\times D \to \mathbf{R}^n$, $R:[0,\infty)\times D\times(0,\epsilon_o]\to\mathbf{R}^n$.

$$f^1(t,x) = \nabla f(t,x)u^1(t,x) - \nabla u^1(t,x)f^o(x),$$

where

$$u^1(t,x) = \int_0^t [f(\tau,x) - f^o(x)]d\tau + a(x),$$

with $a(x)$ a smooth vectorfield such that the average of $u^1$ is zero.

Suppose

a) $f$ has a *Lipschitz*-continuous first derivative in $x$, $g$ and $R$ are *Lipschitz*-continuous in $x$ and all functions are continuous on their domain of definition;

b) $f$ and $g$ are $T$-periodic in $t$, with averages $f^o$ and $g^o$; $|R(t,x,\epsilon)|$ is bounded by a constant uniformly on $[0,\frac{L}{\epsilon})\times D\times(0,\epsilon_o]$;

c) $u(t)$ belongs to an interior subset of $D$ on the time-scale $\frac{1}{\epsilon}$;

then

$$x(t)=u(t)+\epsilon u^1(t,u(t))+O(\epsilon^2)$$

on the time-scale $\frac{1}{\epsilon}$.

**Proof**

First assume $g=0$ and define as in Lemma 3.4.1

$$x(t)=w(t)+\epsilon u^1(t,w(t)).$$

In the periodic case $v(t)=u(t)$, so we have from Lemma's 3.4.2 and 3.4.3

$$w(t)=u(t)+O(\epsilon^2)$$

on the time-scale $\frac{1}{\epsilon}$ so that $x(t)=u(t)+\epsilon u^1(t,(t))+O(\epsilon^2)$ on this time-scale. If $g$ is not zero, we add some $O(\epsilon^2)$ terms involving $g$ in Lemma 3.4.1-3.

### 3.5.2. Example: periodic solutions of the *van der Pol* equation

In § 2.7 we calculated the first order approximation of the *van der Pol*-equation

$$\ddot{x}+x=\epsilon(1-x^2)\dot{x}.$$

For the amplitude $r$ and the phase $\phi$ the equations are

$$\frac{dr}{dt}=\epsilon r\sin^2(t+\phi)[1-r^2\cos^2(t+\phi)],$$

$$\frac{d\phi}{dt}=\epsilon\sin(t+\phi)\cos(t+\phi)[1-r^2\cos^2(t+\phi)].$$

Averaging over $t$ (period $2\pi$) yields

$$\frac{d\tilde{r}}{dt}=\frac{\epsilon}{2}\tilde{r}(1-\frac{\tilde{r}^2}{4}),\quad \frac{d\tilde{\phi}}{dt}=0,$$

producing a periodic solution of the original equation in $r(0)=2$. In the notation of Theorem 3.5.1 we have

$$u^1(t,r,\phi)=\begin{pmatrix} -\frac{r}{2}\sin(2(t+\phi))+\frac{r^3}{32}\sin(4(t+\phi)) \\ -\frac{1}{2}\cos(2(t+\phi))+\frac{r^2}{8}\cos(2(t+\phi))+\frac{r^2}{32}\cos(4(t+\phi)) \end{pmatrix}.$$

For the equation, averaged to second order, we find

$$\frac{d\bar{r}}{dt}=\frac{1}{2}\epsilon\bar{r}(1-\frac{\bar{r}^2}{4}),$$

$$\frac{d\bar{\phi}}{dt}=\frac{\epsilon^2}{8}(-1+\frac{3}{2}\bar{r}^2-\frac{11}{32}\bar{r}^4),$$

where in the notation of Theorem 3.5.1 $u=(\bar{r},\bar{\phi})$. For the periodic solution we find ( with $r(0)=2$, $\phi(0)=0$ ):

$$\bar{r}=2\ ,\ \bar{\phi}\ =\ -\frac{\epsilon^2}{16}t$$

and we have

$$\begin{bmatrix} r(t) \\ \phi(t) \end{bmatrix} = \begin{bmatrix} 2 \\ -\frac{\epsilon^2}{16}t \end{bmatrix} + \epsilon u^1(t,2,-\frac{\epsilon^2}{16}t)+O(\epsilon^2)$$

on the time-scale $\frac{1}{\epsilon}$.

Note that $u^1$, used in this example, has the property that its average over $t$ is zero.

## 3.6. An alternative estimate of first order general averaging

We have by now seen two different approaches to first order averaging: direct estimation of solutions of the differential equations (§ 3) and using a transformation (§ 4); the latter method produces better results, but requires differentiability of the vectorfield. In this section we shall translate the original version of the proof of *Bogoliubov* and *Mitropolsky* in our notation. The original proof is more concerned with continuity in $\epsilon$ than with explicit asymptotic estimates, but the translation is straightforward. The only change here is that we allow for a bounded domain; this introduces some technical difficulties. The idea of the proof is simple: if the transformation function $u^1$ is not differentiable (because the vectorfield is not), we approximate it, using convolution, by a differentiable function $u_\mu^1$. The inherent difficulty is that the gradient of $u_\mu^1$ might be large, in terms of the order of approximation. By a suitable choice of $\mu(\epsilon)$ we find a new proof of Theorem 3.3.3. This proof does not give a sharp estimate in the periodic case, which has therefore to be considered separately.

### 3.6.1. Definition

Let $D^o \subset D$ be such that $dist(\partial D,\partial D^o)>\mu>0$. Let $\psi:D\rightarrow\mathbf{R}^n$ be a continuous vectorfield. Then we define $\psi_\mu:D\rightarrow\mathbf{R}^n$ as

$$\psi_\mu(x)=\int_D \Phi^\mu(x-y)\psi(y)dy,$$

where

$$\Phi^\mu(x)=\begin{cases} A_\mu(1-\frac{\|x\|^2}{\mu^2})\ ,\ \|x\|\leq\mu \\ \quad 0 \quad\ ,\ \|x\|>\mu \end{cases}$$

with $A_\mu\in\mathbf{R}$ such that $\int\Phi^\mu(x)dx=1$.

**3.6.2. Lemma**

$$\int \nabla \Phi^{\mu}(x)dx = O(\frac{1}{\mu}).$$

**Proof**

Straightforward computation. □

**3.6.3. Lemma**

If $\psi: D \rightarrow \mathbf{R}^n$ is uniformly bounded and *Lipschitz*-continuous, i.e. $\|\psi(x)-\psi(y)\| \leqslant \lambda \|x-y\|$ for all $x,y \in D$, then

$$\psi(x) = \psi_{\mu}(x) + O(\mu),$$

uniformly on $D^o$.

**Proof**

Let $x \in D^o$. Then

$$\|\psi(x)-\psi_{\mu}(x)\| = \|\psi(x) - \int_D \Phi^{\mu}(x-y)\psi(y)dy\|$$

$$= \|\psi(x)\int \Phi^{\mu}(x-y)dy - \int \Phi^{\mu}(x-y)\psi(y)dy\|$$

$$= \|\int \Phi^{\mu}(x-y)(\psi(x)-\psi(y))dy\|$$

$$\leqslant \int |\Phi^{\mu}(x-y)| \lambda \|x-y\| dy \leqslant \lambda\mu.$$

□

**3.6.4. Lemma**

Suppose $f$ is *Lipschitz*-continuous with respect to $x$. Let $x(t)$ be the solution of

$$\dot{x} = \epsilon f(t,x)\ ,\ x(0) = \xi.$$

Let $w(t)$ be defined by

$$x(t) = w(t) + \delta_1(\epsilon)u^1(t,w(t)),$$

where

$$\delta_1(\epsilon)u^1(t,w) = \epsilon \int_0^t [f(\tau,w) - f^o(w)]d\tau.$$

Then

$$w(t) = \xi + \epsilon \int_0^t f^o(w(\tau))d\tau + O(\delta_1^{1/2}(\epsilon)).$$

**Proof**

$$w(t)=x(t)-\delta_1(\epsilon)u^1(t,w(t))$$

$$=\xi+\epsilon\int_0^t f(\tau,x(\tau))d\tau-\delta_1(\epsilon)u_\mu^1(t,w(t))+O(\delta_1(\epsilon)\mu)$$

$$=\xi+\epsilon\int_0^t f(\tau,x(\tau))d\tau-\delta_1(\epsilon)\int_0^t\frac{d}{d\tau}u_\mu^1(\tau,w(\tau))d\tau+O(\delta_1(\epsilon)\mu)$$

$$=\xi+\epsilon\int_0^t f(\tau,x(\tau))d\tau-\epsilon\int_0^t[f_\mu(\tau,w(\tau))-f_\mu^o(w(\tau))]d\tau$$

$$+\epsilon\int_0^t[f(\tau,x(\tau))-f_\mu(\tau,x(\tau))]d\tau$$

$$+\delta_1(\epsilon)\int_0^t[\epsilon\nabla f_\mu u_\mu^1-\nabla u_\mu^1\frac{dw}{d\tau}]d\tau+O(\delta_1(\epsilon)\mu)$$

$$=\xi+\epsilon\int_0^t f(\tau,x(\tau))d\tau+O(\mu)+O(\frac{\delta_1(\epsilon)}{\mu}).$$

So we let $\mu=\delta_1^{1/2}(\epsilon)$ and we obtain the desired estimate.□

Using this Lemma and *Gronwall*'s Lemma, we obtain a second proof of Theorem 3.3.3.

### 3.6.5. Remark

If one surveys the proof of Lemma 2.6.4, one gets the distinct feeling that sharper estimates must be possible. One could for instance apply the averaging theory to the function $g_\mu(t,x)=f(t,x)-f_\mu(t,x)$. The difficulty here is, that while $g_\mu$ is itself $O(\mu)$, its *Lipschitz*-constant can only be shown to be $O(1)$; it is also not clear that the $\delta$-function belonging to $g_\mu$, is smaller than $\delta_1$, even if this might seem intuitively acceptable. A problem here is that $\dot{x}=O(\epsilon)$ and not $O(\epsilon\mu)$.

## 3.7. Application of general averaging to almost-periodic vectorfields

In this section we discuss some questions that arise in studying initial value problems of the form

$$\frac{dx}{dt}=\epsilon f(t,x)\ ,\ x(0)=x_o,$$

with $f$ almost-periodic in $t$. For the basic theory of almost-periodic functions we refer to an introduction by *Bohr* (Boh32a). A more recent introduction with the emphasis on its use in differential equations has been given by *Fink* (Fin74a). In this book averaging has been discussed for proving existence of almost-periodic solutions. Both qualitative and quantitative aspects of almost-periodic solutions of periodic and almost-periodic differential equations has been given extensive treatment by *Roseau* (Ros70a).

A simple example of an almost-periodic function is found by taking the sum of two periodic functions like

$$f(t)=\sin(t)+\sin(2\pi t).$$

Several equivalent definitions are in use; we take a two-step definition by *Bohr*.

**3.7.1. Definition**

A subset $S$ of $\mathbf{R}$ is called *relatively dense* if there exists a positive number $L$ such that $[a,a+L]\cap S\neq\varnothing$ for all $a\in\mathbf{R}$. The number $L$ is called the inclusion length.

**3.7.2. Definition**

1) Consider a vectorfield $f(t)$, continuous on $\mathbf{R}$, and a positive number $\epsilon$; $\tau(\epsilon)$ is a *translation-number* of $f$ if

$$\|f(t+t(\epsilon))-f(t)\|\leqslant\epsilon \textit{ for all } t\in\mathbf{R};$$

2) The vectorfield $f(t)$, continuous on $\mathbf{R}$, is called *almost-periodic* if for each $\epsilon>0$ a *relatively dense* set of *translation-numbers* $\tau(\epsilon)$ exists.

In the context of averaging the following result is basic.

**3.7.3. Lemma**

Consider the continuous vectorfield $f:\mathbf{R}\times\mathbf{R}^n\to\mathbf{R}^n$; if $f(t,x)$ is almost-periodic in $t$ and *Lipschitz*-continuous in $x$, $f$ is a *KBM*-vectorfield.

**Proof**

This is a trivial generalization of § 50 in *Bohr* (Boh32a). □

It follows immediately that with the appropriate assumptions of Theorem 3.3.3 or 3.4.5 we can apply general averaging to the almost-periodic differential equation. Suppose the conditions of Theorem 3.4.5 have been satisfied, then introducing again the averaged equation

$$\frac{dy}{dt}=\epsilon f^o(y)\ ,\ y(0)=x_o,$$

we have

$$x(t)=y(t)+O(\delta_{(}\epsilon))$$

on the time-scale $\frac{1}{\epsilon}$, where

$$\delta_1(\epsilon)=\sup_{x\in D}\ \sup_{t\in[0,\frac{L}{\epsilon})}\epsilon\left|\int_0^t[f(\tau,x)-f^o(x)]d\tau\right|.$$

We shall discuss the magnitude of the error $\delta_1(\epsilon)$. Many cases in practice are covered by the following lemma:

### 3.7.4. Lemma

If we can decompose the almost-periodic vectorfield $f(t,x)$ as a finite sum of $N$ periodic vectorfields

$$f(t,x)=\sum_{n=1}^{N} f_n(t,x),$$

we have $\delta_1(\epsilon)=\epsilon$ and, moreover,

$$x(t)=y(t)+O(\epsilon).$$

**Proof**

Interchanging the finite summation and the integration gives the desired result. □

A fundamental obstruction against generalizing this Lemma is that in general

$$\int_o^t [f(t,x)-f^o(x)]dt$$

with $f$ almost-periodic, need not be bounded (see the example below). One might be tempted to apply the approximation theorem for almost-periodic functions: For each $\epsilon>0$ we can find $N(\epsilon)\in\mathbf{N}$ such that

$$\|f(t,x)-\sum_{n=1}^{N} a_n(x)e^{i\lambda_n t}\|\leqslant\epsilon$$

for $t\in[0,\infty)$ and $\lambda_n\in\mathbf{R}$ (Cf. (Boh32a), § 84).

In general however, $N$ depends on $\epsilon$, which destroys the possibility of obtaining an $O(\epsilon)$ estimate for $\delta_1$. The difficulties can be illustrated by the following initial value problem.

### 3.7.5. Example

Consider the equation

$$\frac{dx}{dt}=\epsilon a(x)f(t)\ ,\ x(0)=x_o.$$

The function $a(x)$ is sufficiently smooth in some domain $D$ containing $x_o$. We define the almost-periodic function

$$f(t)=\sum_{n=1}^{\infty}\frac{1}{2^n}\cos(\frac{t}{2^n}).$$

Note that this is a uniformly convergent series consisting of continuous terms so we may integrate $f(t)$ and interchange summation and integration.

For the average of the right hand side we find

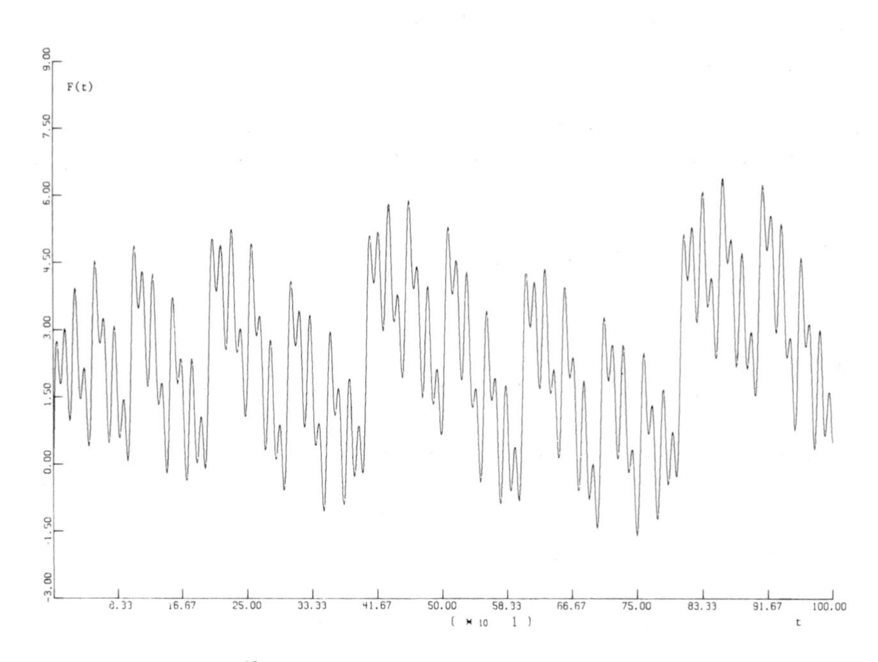


**Figure 3.7-1** $F(t) = \sum_{n=1}^{\infty} \sin(\frac{t}{2^n})$ *as a function of time on* $[0,10000]$. *The function F is the integral of an almost-periodic function and is not uniformly bounded.*

$$\lim_{T\to\infty} \frac{1}{T} a(x) \int_0^T f(t)dt = 0.$$

So we have

$$x(t) = x_o + O(\delta_1(\epsilon))$$

on the time-scale $\frac{1}{\epsilon}$. Suppose $\sup_{x \in D} a(x) = M$ then we have

$$\delta_1(\epsilon) = \sup_{t\in[0,\frac{L}{\epsilon}]} \epsilon M \left| \sum_{n=1}^{\infty} \sin(\frac{t}{2^n}) \right| .$$

It is easy to see that as $\epsilon \to 0$, $\frac{\delta_1(\epsilon)}{\epsilon}$ becomes unbounded, so the error in this almost-periodic case is larger than $O(\epsilon)$. In figure 3.7-1 we illustrate this behavior of $\delta_1(\epsilon)$ and $F(t) = \sum_{n=1}^{\infty} \sin(\frac{t}{2^n})$. A simple example is in the case $a(x) = 1$; we have explicitly

$$x(t) = x_o + \epsilon F(t).$$

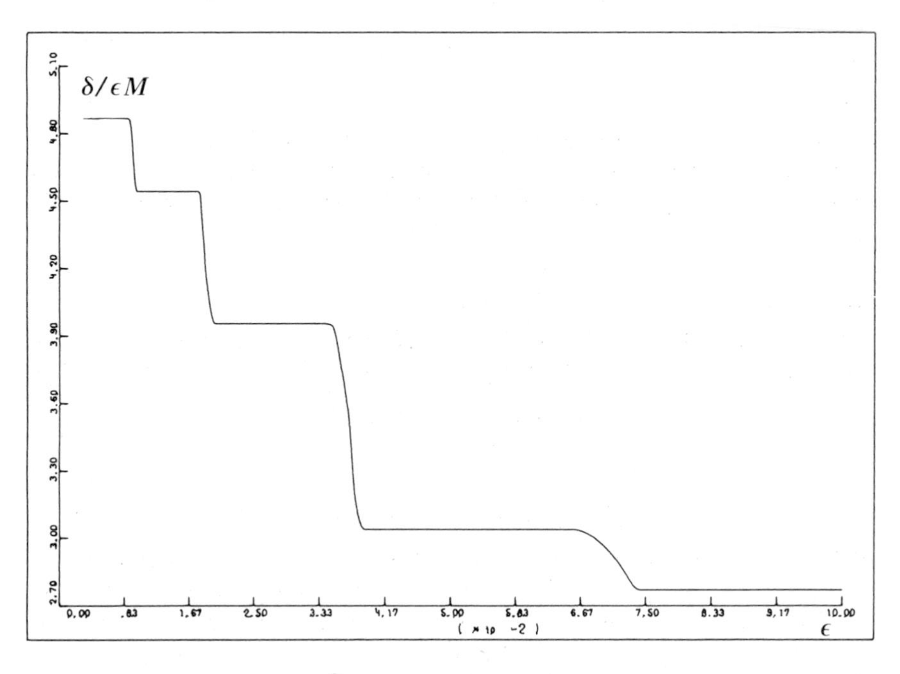


**Figure 3.7-2** *The quantity* $\frac{\delta_1}{\epsilon M}$ *as a function of* $\epsilon$ *obtained from the analysis of* $F(t)$. *As* $\epsilon$ *decreases,* $\frac{\delta_1}{\epsilon M} = \sup_{0 \leq \epsilon t \leq 1} F(t)$ *increases.*

The same type of error arises if the solutions are bounded. Take for example $a(x)=x(1-x)$, $0<x_o<1$. We find

$$x(t)=\frac{x_o e^{\epsilon F(t)}}{1-x_o+x_o e^{\epsilon F(t)}}.$$

Sometimes an $O(\epsilon)$-estimate can be obtained by studying the generalized *Fourier*-expansion of an almost-periodic vectorfield

$$f(t,x)=f^o(x)+\sum_{n=1}^{\infty}[a_n(x)\cos(\lambda_n t)+b_n(x)\sin(\lambda_n t)]$$

with $\lambda_n>0$. We have:

### 3.7.6. Lemma

Suppose the conditions of Theorem 3.4.5 have been satisfied for the initial value problems

$$\frac{dx}{dt}=\epsilon f(t,x)\ ,\ x(0)=x_o$$

and

$$\frac{dy}{dt}=\epsilon f^o(y)\ ,\ y(0)=x_o.$$

If $f(t,x)$ is an almost-periodic vectorfield with a generalized *Fourier*-expansion such that $\lambda_n \geqslant \alpha > 0$ with $\alpha$ independent of $n$ then

$$x(t) = y(t) + O(\epsilon)$$

on the time-scale $\frac{1}{\epsilon}$.

**Proof**

If $\lambda_n \geqslant \alpha > 0$ we have that

$$I(t,x) = \int_0^t [f(\tau,x) - f^o(x)]d\tau$$

is an almost-periodic vectorfield; See (Fin74a), chapter 4.8. So $|I(t,x)|$ is bounded for $t \geqslant 0$ which implies $\delta_1(\epsilon) = O(\epsilon)$.□

## 3.8. Application to systems, containing terms slowly varying with time

Many technical and physical applications of the theory of asymptotic approximations concerns problems in which certain quantities exhibit slow variation with time. Consider for instance a pendulum with variable length, a spring with varying stiffness or a mechanical system from which mass is lost.

From the point of view of the theory of averaging the application of the preceding theory is simple (unless there are passage through resonance problems, cf. chapter 5), the technical obstructions however can be considerable. We illustrate this as follows. Suppose the system has been put in the form

$$\frac{dx}{dt} = \epsilon f(t, \epsilon t, x)\ ,\ x(0) = x_o,$$

with $x \in \mathbf{R}^n$. Introduce the new independent variable

$$\tau = \epsilon t.$$

Then we have the $(n+1)$-dimensional system in the standard form

$$\frac{dx}{dt} = \epsilon f(t,\tau,x)\ ,\ x(0) = x_o,$$

$$\frac{d\tau}{dt} = \epsilon \qquad ,\ t(0) = 0.$$

Suppose we may average the vectorfield over $t$, then an approximation can be obtained by solving the initial value problem

$$\frac{dy}{dt} = \epsilon f^o(\tau,y)\ ,\ y(0) = x_o,$$

$$\frac{d\tau}{dt} = \epsilon \qquad ,\ \tau(0) = 0.$$

or

$$\frac{dy}{dt} = \epsilon f^o(\epsilon t, y) \ , \ y(0) = x_o.$$

So the recipe is simply: average over $t$, keeping $\epsilon t$ and $x$ fixed, and solve the resulting equation. In practice this is not always so easy; we consider a simple example below. *Mitropolsky* devoted a book (Mit65a) to the subject with many more details and examples. Some problems with slowly varying time in celestial mechanics are considered in appendix 8.7.

### 3.8.1. Example

Consider a pendulum with slowly varying length and some other perturbations to be specified later on. If we put the mass and the gravitational constant equal to one, and if we put $l = l(\epsilon t)$ for the length of the pendulum, we have according to (Mit65a) the equation

$$\frac{d}{dt}(l^2(\epsilon t)\dot{x}) + l(\epsilon t)x = g(\epsilon t, x, \dot{x}),$$

with initial values given. The first problem is to put this equation in standard form. If $\epsilon = 0$ we have a harmonic oscillator with frequency $\omega_o = l(0)^{-\frac{1}{2}}$ and solutions of the form $\cos(\omega_o t + \phi)$. This inspires us to introduce another time-like variable

$$s = \int_0^t l(\epsilon\sigma)^{-\frac{1}{2}} d\sigma.$$

If $\epsilon = 0$, $s$ reduces to the natural time-like variable $\omega_o t$. For $s$ to be time-like we require $l(\epsilon t)$ to be such that $s(t)$ increases monotonically and that $t \to \infty \Rightarrow s \to \infty$. We abbreviate $\epsilon t = \tau$; note that since

$$\frac{d}{dt} = l^{-\frac{1}{2}}(\tau)\frac{d}{ds}$$

the equation becomes

$$\frac{d^2x}{ds^2} + x = \epsilon l^{-1}(\tau) g(\tau, x, l^{-\frac{1}{2}}(\tau)\frac{dx}{ds}) - \frac{3}{2}\epsilon l^{-\frac{1}{2}}(\tau)\frac{dl}{d\tau}\frac{dx}{ds}.$$

Introducing phase-amplitude coordinates by

$$x = r\cos(s+\phi),$$

$$\frac{dx}{ds} = -r\sin(s+\phi),$$

produces the standard form

$$\frac{dr}{ds} = -\epsilon\sin(s+\phi)l^{-1}(\tau)g(\tau, x, l(\tau)^{-\frac{1}{2}}\frac{dx}{ds}) - \frac{3}{2}\epsilon l(\tau)^{-\frac{1}{2}}\frac{dl}{d\tau}r\sin^2(s+\phi),$$

$$\frac{d\phi}{ds} = -\frac{\epsilon}{r}\cos(s+\phi)l^{-1}(\tau)g(\tau, x, l(\tau)^{-\frac{1}{2}}\frac{dx}{ds}) - \frac{3}{4}\epsilon l(\tau)^{-\frac{1}{2}}\frac{dl}{d\tau}\sin(2(s+\phi)),$$

$$\frac{d\tau}{ds} = \epsilon l^{\frac{1}{2}}(\tau).$$

Initial values $r(0)=r_o$, $\phi(0)=\phi_o$, and $\tau(0)=0$. Averaging over $s$ does not touch the last equation, so we still have $\tau=\epsilon t$, as it should be. We consider two cases.

### 3.8.1.1. The linear case, $g=0$

Averaging produces

$$\frac{d\tilde{r}}{ds}=\frac{3}{4}\epsilon\tilde{r}l^{-1/2}\frac{dl}{d\tau},$$

$$\frac{d\tilde{\phi}}{ds}=0.$$

The first equation can be written as

$$\frac{d\tilde{r}}{d\tau}=-\frac{3}{4}\tilde{r}l^{-1}\frac{dl}{d\tau},$$

so we have with Theorem 3.2.10 on the time-scale $\frac{1}{\epsilon}$ in $s$

$$r(\tau)=r_o\left(\frac{l(0)}{l(\tau)}\right)^{\frac{3}{4}}+O(\epsilon)\ ,\ \phi(\tau)=\phi_o+O(\epsilon).$$

### 3.8.1.2. A nonlinear perturbation with damping

Suppose that the oscillator has been derived from the mathematical pendulum so that we have a *Duffing*-type of perturbation (coefficient $\mu$); moreover we have small linear damping (coefficient $\sigma$). We put

$$g=\mu l(\epsilon t)x^3-\sigma l(\epsilon t)\dot{x}.$$

The standard form for $r$ and $\phi$ becomes

$$\frac{dr}{ds}=-\epsilon\mu r^3\sin(s+\phi)\cos^3(s+\phi)-\epsilon\sigma l^{-1/2}r\sin(s+\phi)-\frac{3}{2}\epsilon l^{-1/2}\frac{dl}{d\tau}r\sin^2(s+\phi),$$

$$\frac{d\phi}{ds}=-\epsilon\mu r^2\cos^4(s+\phi)-1/2\epsilon\sigma l^{-1/2}\sin(2(s+\phi))-\frac{3}{4}\epsilon l^{-1/2}\frac{dl}{d\tau}\sin(2(s+\phi)).$$

Averaging produces for $r$ and $\phi$

$$\frac{d\tilde{r}}{ds}=-\epsilon l^{-1/2}\frac{\tilde{r}}{2}\left(\sigma+\frac{3}{2}\frac{dl}{d\tau}\right),$$

$$\frac{d\tilde{\phi}}{ds}=-\frac{3}{8}\epsilon\mu\tilde{r}^2.$$

If $\tilde{r}$ is known, $\tilde{\phi}$ can be found by direct integration. The equation for $\tilde{r}$ can be written as

$$\frac{d\tilde{r}}{d\tau}=-l^{-1}\frac{\tilde{r}}{2}\left(\sigma+\frac{3}{2}\frac{dl}{d\tau}\right),$$

so we have with Theorem 3.2.10 on the time-scale $\frac{1}{\epsilon}$ in $s$

$$r(\tau)=r_o\left(\frac{l(0)}{l(\tau)}\right)^{\frac{3}{4}} e^{-\frac{\sigma}{2}\int_0^{\tau} l^{-1}(u)du} + O(\epsilon).$$

**Remark**

Some interesting studies have been devoted to the equation

$$\ddot{y}+\omega^2(\epsilon t)y=0.$$

The relation with our example becomes clear when we put $g=0$ and transform $x=\frac{y}{l}$. If $l$ can be differentiated twice, we find

$$\ddot{y}+l^{-1}(1-\ddot{l})y=0.$$

## 3.9. Approximations on the time-scale $1/\epsilon^2$ in the periodic case

The asymptotic approximations obtained thus far, are valid on the time-scale $\frac{1}{\epsilon}$. One may wonder whether sometimes results on longer time-scales can be obtained, although we are aware that the general theory (in its present form, at least) can not be extended to longer time-scales (Cf. the counter examples in § 2.7). One natural context for extension of the time-scale of validity is in problems with attraction (See the next chapter). Here we shall treat another natural extension which applies frequently in practice.

Consider the initial value problem

$$\dot{x}=\epsilon f(t,x)+\epsilon^2 g(t,x)+\epsilon^3 R(t,x,\epsilon)\ ,\ x(0)=x_o.$$

Suppose that the conditions of Theorem 3.4.5 are satisfied and that moreover

$$f^o=0.$$

Clearly, $x(t)=x_o+O(\epsilon)$ on the time-scale $\frac{1}{\epsilon}$. We shall show that $u(t)$, the solution of

$$\dot{u}=\epsilon^2 f^{1o}(u)+\epsilon^2 g^o(u)\ ,\ u(0)=x_o$$

is an $O(\epsilon)$ approximation of $x(t)$ on the time-scale $\frac{1}{\epsilon^2}$. The construction and proof runs as follows.

Define $w(t)$ by

$$x(t)=w(t)+\epsilon u^1(t,w(t)),$$

with

$$u^1(t,w)=\int_0^t f(\tau,w)d\tau-\frac{1}{T}\int_0^T\int_0^t f(\tau,w)d\tau dt.$$

($f$ and $g$ are $T$-periodic in $t$). Substitution in the differential equation produces for $w$

$$\dot{w}+\epsilon\nabla u^1(t,w)\dot{w}+\epsilon f(t,w)$$
$$=\epsilon f(t,w+\epsilon u^1(t,w))+\epsilon^2 g(t,w+\epsilon u^1(t,w))+O(\epsilon^3);$$

$w(0)$ is determined by the equation

$$x_o=w(0)+\epsilon u^1(0,w(0))$$

or

$$w(0)=x_o-\epsilon u^1(0,x_o)+O(\epsilon^2).$$

Expanding the right hand side of the differential equation we have

$$(I+\epsilon\nabla u^1(t,w))\dot{w}=\epsilon^2\nabla f(t,w)u^1(t,w)+\epsilon^2 g(t,w)+O(\epsilon^3),$$

with $I$ the $n\times n$ unit matrix. $\nabla u^1$ is a bounded matrix so we can invert and expand to obtain

$$\dot{w}=\epsilon^2\nabla f(t,w)u^1(t,w)+\epsilon^2 g(t,w)+O(\epsilon^3),$$

with $w(0)$ determined above. We estimate $x(t)-u(t)$ with the triangle inequality

$$\|x(t)-u(t)\|\leqslant\|x(t)-w(t)\|+\|w(t)-u(t)\|;$$

$\|x(t)-w(t)\|=O(\epsilon)$ as long as $x(t)$ and $w(t)$ remain in $D$. $\|w(t)-u(t)\|$ can be estimated with Theorem 2.6.3; in the notation of this Theorem we have $\delta_1(\epsilon)=\epsilon^2$, $\delta_2(\epsilon)=\epsilon^3$ and $\delta_3(\epsilon)=\epsilon$. So

$$\|w(t)-u(t)\|=O(\epsilon)$$

on the time-scale $\frac{1}{\epsilon^2}$. This completes the proof of the next Theorem.

**3.9.1. Theorem (first order approximation on the time-scale $\frac{1}{\epsilon^2}$ (Bur74a))**

Consider the initial value problems

$$\dot{x}=\epsilon f(t,x)+\epsilon^2 g(t,x)+\epsilon^3 R(t,x)\ ,\ x(0)=x_o$$

and

$$\dot{u}=\epsilon f^{1o}(u)+\epsilon^2 g^o(u)\ ,\ u(0)=x_o,$$

with $f,g{:}\mathbf{R}\times\mathbf{R}^n\to\mathbf{R}^n$, $R{:}\mathbf{R}\times\mathbf{R}^n\times(0,\epsilon_o]\to\mathbf{R}^n$, $x,u,x_o\in D\subset\mathbf{R}^n$, $t\in[0,\infty)$ and $\epsilon\in(0,\epsilon_o]$; As usual

$$f^1(t,x)=\nabla f(t,x)u^1(t,x)$$

and

$$u^1(t,x)=\int_0^t f(\tau,x)d\tau-\frac{1}{T}\int_0^T\int_0^t f(\tau,x)d\tau dt.$$

Suppose

a) $f$ has a *Lipschitz*-continuous first derivative in $x$, $g$ and $R$ are *Lipschitz*-continuous in $x$ on $D$; $f,g,R$ are continuous in $t$ and $x$ with $t$ on the time-scale $\frac{1}{\epsilon^2}$;

b) $f$ and $g$ are $T$-periodic in $t$, averages $f^o$ and $g^o$ ($f^1$ has average $f^{1o}$); $R$ is bounded by a constant independent of $\epsilon$ for $x \in D$, $t$ on the time-scale $\frac{1}{\epsilon^2}$. Moreover $f^o = 0$;

c) $u(t)$ belongs to an interior subset of $D$ on the time-scale $\frac{1}{\epsilon^2}$;

then

$$x(t) = u(t) + O(\epsilon)$$

on the time-scale $\frac{1}{\epsilon^2}$.

**3.9.2. Remark**

Note that we had only a priori knowledge of the existence of $x$ on the time-scale $\frac{1}{\epsilon}$, so this is not only an approximation result, but also an existence theorem. Of course condition (c) is important here.

**3.9.3. Remark**

One may wonder if much work is needed in practice to extend the results to obtain an $O(\epsilon^2)$ approximation on the time-scale $\frac{1}{\epsilon^2}$. The answer is yes. In the proof we approximated $w$ by $u$, neglecting $O(\epsilon^3)$ terms. For an $O(\epsilon^2)$ approximation these terms have to be computed and included in the averaged equation.

**3.9.4. Remark**

An important point concerns the choice of $u^1$. Inspection of the proof shows that we may as well put

$$u^1(t,x) = \int_0^t f(\tau,x)d\tau + h(x),$$

with $h$ sufficiently smooth. The result of the Theorem remains the same; however for the particular choice of $h$ that we took, the computational effort has been minimized. Another advantage is that this choice of $h$ also facilitates the calculation of $O(\epsilon^2)$ approximations on the time-scale $\frac{1}{\epsilon^2}$.

**3.9.5. Remark**

We may extend Theorem 3.9.1 to the case of general averaging using the same type of proof and the formulation of Theorem 3.4.6. This will be left as an exercise for the reader.

**3.9.6. Example**

Consider the modified *van der Pol*-equation

$$\ddot{x}+x-\epsilon x^2=\epsilon^2(1-x^2)\dot{x}.$$

We choose an amplitude-phase representation to obtain perturbation equations in the standard form: $(x,\dot{x})\mapsto(r,\phi)$ by $x=r\cos(t+\phi)$, $\dot{x}=-r\sin(t+\phi)$. We find

$$\frac{dr}{dt}=-\epsilon r^2\sin(t+\phi)\cos^2(t+\phi)+\epsilon^2\sin^2(t+\phi)[1-r^2\cos^2(t+\phi)],$$

$$\frac{d\phi}{dt}=-\epsilon r\cos^3(t+\phi)+\epsilon^2\cos(t+\phi)[1-r^2\cos^2(t+\phi)]\sin(t+\phi).$$

Conditions (a) and (b) of Theorem 3.9.1 have been satisfied.

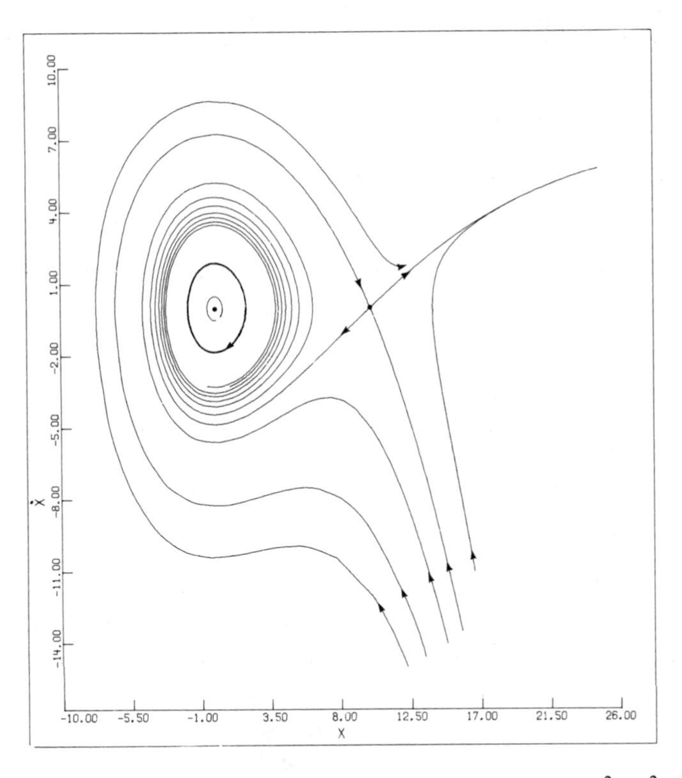


**Figure 3.9-1** *The $x,\dot{x}$-phase plane of the equation* $\ddot{x}+x-\epsilon x^2=\epsilon^2(1-x^2)\dot{x}$; $\epsilon=0.1$.

The averaged equations describe the flow on the time-scale $\frac{1}{\epsilon^2}$ with error $O(\epsilon)$. The saddle point behavior has not been described by this perturbation approach as the saddle point coordinates are $(\frac{1}{\epsilon},0)$.

We put

$$u^1(t,r,\phi)=\begin{bmatrix} \frac{1}{3}r^2\cos^3(t+\phi) \\ -r\sin(t+\phi)+\frac{1}{3}r\sin^3(t+\phi) \end{bmatrix}.$$

After the calculation of $\nabla f u^1$ and averaging we find

$$\frac{d\tilde{r}}{dt}=\tfrac{1}{2}\epsilon^2\tilde{r}(1-\frac{\tilde{r}^2}{4}),$$

$$\frac{d\tilde{\phi}}{dt}=-\epsilon^2\frac{5}{12}\tilde{r}^2.$$

We conclude that as in the *van der Pol*-equation we have a stable periodic solution with amplitude $r=2+O(\epsilon)$ (Cf. § 2.7). The $O(\epsilon)$ term in the original equation only induces a shifting of the phase-angle $\phi$. For the periodic solution we have

$$x(t)=2\cos(t-\frac{5}{3}\epsilon^2 t)+O(\epsilon)$$

on the time-scale $\frac{1}{\epsilon^2}$. See Figure 3.9-1 for the phase-portrait.

Important examples of Theorem 3.9.1 in the theory of *Hamiltonian* systems will be treated later on (Chapter 7).

# 4. Attraction

## 4.1. Some examples

Solutions of differential equations can be attracting to a particular solution and this phenomenon may assume many different forms. Suppose for instance we consider an initial value problem in $\mathbf{R}^n$ of the form

$$\dot{x} = Ax + \epsilon f(t,x) \ , \ x(0) = x_o.$$

The matrix $A$ is constant and all the eigenvalues have negative real parts. If $\epsilon=0, x=0$ is an attractor ; how do we approximate the solution if $\epsilon \neq 0$ and what are the conditions to obtain an approximation of the solution on the time-scale $\frac{1}{\epsilon}$ or even $[0,\infty)$? Also we should like to extend the problem to the case

$$\dot{x} = Ax + g(x) + \epsilon f(t,x) \ , \ x(0)=x_o,$$

where we suppose that the equation with $\epsilon=0$ has $x=0$ as an attracting solution with domain of attraction $D$ . How do we obtain an approximation of the solution if $x_o \in D$ and $\epsilon \neq o$ ?

### 4.1.1. Example

Consider the problem, encountered in § 2.3, of two species with a restricted supply of food and a slight negative interaction between the species. The growth of the population densities $x_1$ and $x_2$ can be described by the system

$$\dot{x}_1 = ax_1 - bx_1^2 - \epsilon x_1 x_2,$$

$$\dot{x}_2 = cx_2 - dx_2^2 - \epsilon e x_1 x_2,$$

where $a,b,c,d,e$ are positive constants. Putting $\epsilon=0$ one notes that $(\frac{a}{b},\frac{c}{d})$ is a positive attractor; The domain of attraction D is given by $x_1>0, x_2>0$. In figure 4.1-1 we give an example of the phase-plane with $\epsilon\neq0$ and $\epsilon=0$.

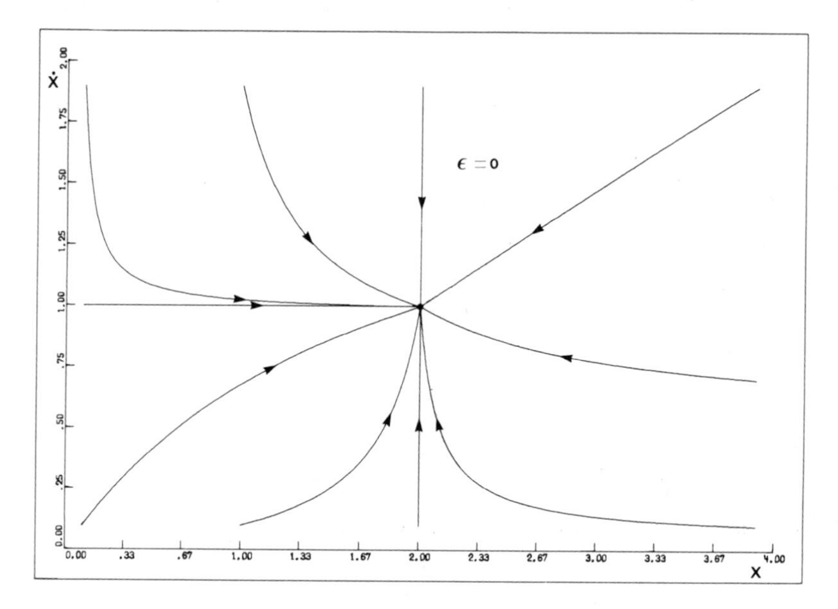


**Figure 4.1-1a** *Phase-plane for the system*

$$\dot{x}_1 = x_1 - \tfrac{1}{2}x_1^2 - \epsilon x_1 x_2$$

$$\dot{x}_2 = x_2 - x_2^2 - \epsilon x_1 x_2$$

*for* $\epsilon=0$, *i.e without interaction of the species.*

Another attraction problem arises in the following way. Consider the problem

$$\dot{x} = \epsilon f(t,x) \quad , \quad x(0)=x_o.$$

Suppose that we may average and that the equation

$$\dot{x} = \epsilon f^o(x)$$

contains an attractor, domain of attraction $D$ . Can we extend the time-scale of validity of the approximation if we start the solution in $D$ ?

### 4.1.2. Example

Anharmonic oscillator with linear damping:

$$\ddot{x} + x = -\epsilon\dot{x} + \epsilon x^3.$$

Putting $x=r\cos(t+\psi), \dot{x}=-r\sin(t+\psi)$ we obtain (Cf. § 2.4)

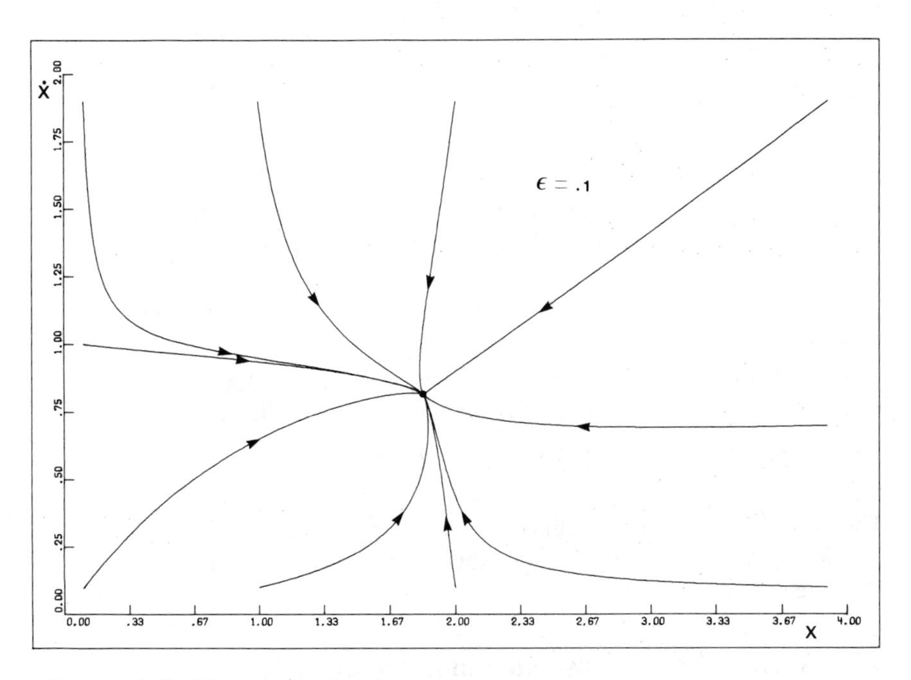


**Figure 4.1-1b** *Phase-plane for the system*

$$\dot{x}_1 = x_1 - \tfrac{1}{2}x_1^2 - \epsilon x_1 x_2$$

$$\dot{x}_2 = x_2 - x_2^2 - \epsilon x_1 x_2$$

*for* $\epsilon = 0.1$, *i.e. with (small) interaction of the species.*

$$\dot{r} = -\epsilon r \sin^2(t+\psi) - \epsilon r^3 \sin(t+\psi)\cos^3(t+\psi),$$

$$\dot{\psi} = -\epsilon \sin(t+\psi)\cos(t+\psi) - \epsilon r^2 \cos^4(t+\psi),$$

or upon averaging over $t$

$$\dot{\tilde{r}} = -\frac{\epsilon}{2}\tilde{r},$$

$$\dot{\tilde{\psi}} = -\frac{3}{8}\epsilon \tilde{r}^2,$$

Note that we have no isolated critical point (at least not in polar coordinates) of the averaged system which is attracting. This is a general feature of the harmonic oscillator with an autonomous perturbation (Cf. § 2.6, where we analyze the *van der Pol*-equation). One can obtain an attractor in a simple way by the following reduction: put $t+\psi=\theta$ in the equations for $r$ and $\psi$. For $\theta$ we have

$$\dot{\theta} = 1 - \epsilon \sin\theta\cos\theta - \epsilon r^2 \cos^4\theta$$

so that

$$\dot{r} = -\epsilon r \sin^2\theta - \epsilon r^3 \sin\theta\cos^3\theta.$$

Averaging over $\theta$ produces

$$\frac{d\tilde{r}}{d\theta} = -\tfrac{1}{2}\epsilon\tilde{r}.$$

This is a first order equation with attractor $\tilde{r}=0$. Can we extend the time-scale of validity of the approximation in $\theta$? A second question which came up in the analysis of this example is the following. Suppose we have 'attraction in one or more' but not in all components of the averaged vector field. Can we extend the time-scale of validity in the original independent variable $t$?

The ideas presented in this chapter have been around for some time. We mention the papers of *Banfi* and *Graffi* (Ban67a) and (Ban69a).

More detailed proofs were given by *Balachandra* and *Sethna* (Bal75a) and *Eckhaus* (Eck75a).

The proofs can be simplified a little by using a lemma due to *Sanchez − Palencia* (San75a) and this is the approach which we shall use in this chapter.

## 4.2. Averaged equations containing an attractor

Consider the following differential equation

$$\dot{x} = \epsilon f(t,x),$$

with $x \in D \subset \mathbf{R}^n$. Suppose that $f$ is a *KBM*-vectorfield. The averaged equation is

$$\dot{y} = \epsilon f^o(y).$$

We know from the averaging theorems in the preceding chapter that if we supply these equations with an initial value $x_o \in D^o \subset D$ , the solutions stay $\delta(\epsilon)$ -close on the time-scale $\frac{1}{\epsilon}$ ; here $\delta(\epsilon)=o(1)$. Suppose now that

$$f^o(0) = 0$$

and that $y=0$ is an attractor for all the solutions $y(t)$ starting in $D^o$ (if this statement holds for $y=x_c$ with $f^o(x_c)=0$ , we translate this critical point to the origin). In fact we suppose somewhat more: we can write

$$f^o(y) = Ay + g(y),$$

with $g(0)=0$ and $A$ a constant $n \times n$-matrix with the eigenvalues having negative real parts only. The matrix $A$ *does not* depend on $\epsilon$ ; if it does some special problems may arise, see *Robinson* (Rob83a) and § 6 of this chapter. The vectorfield $g$ represents the nonlinear part of $f^o$ near $y=0$.

We have seen in Ch. 1 that the *Poincaré-Lyapunov* theorem guarantees that the solutions attract exponentially towards the origin. Starting in the *Poincaré-Lyapunov* neighborhood of the origin we have

$$\|x(t)\| \leqslant C\|x_o\|e^{-\mu t},$$

with $C$ and $\mu$ positive constants. Moreover we have for two solutions $x_1(t)$ and $x_2(t)$ starting in a neighborhood of the origin that from a certain time $t = t_o$ onwards

$$\|x_1(t) - x_2(t)\| \leqslant C \|x_1(t_o) - x_2(t_o)\| e^{-\mu(t-t_o)}.$$

See theorem 1.4.4 and lemma 1.4.6. We shall now apply these results together with averaging to obtain asymptotic approximations on $[0, \infty)$. Starting outside the *Poincaré-Lyapunov* domain, averaging provides us with a time-scale $\frac{1}{\epsilon}$ which is long enough to reach the domain $\|x\| \leqslant \delta$ where exponential contraction takes place. A summation trick, in this context proposed by *Sanchez – Palencia* (San75a) will take care of the growth of the error on $[0, \infty)$. A different proof has been given by *Eckhaus* (Eck75a); see also *Sanchez – Palencia* (San76a) where the method is placed in the context of Banach spaces and where one can also find a discussion of the perturbation of orbits in phase-space.

**4.2.1. Theorem** (*Eckhaus / Sanchez – Palencia*)

Consider the initial value problem

$$\dot{x} = \epsilon f(t, x) \ , \ x(0) = x_o,$$

with $x_o, x \in D \subset \mathbf{R}^n$. Suppose $f$ is a *KBM*-vectorfield producing the averaged equation

$$\dot{y} = \epsilon f^o(y) \ , \ y(0) = x_o,$$

where $y = 0$ is an asymptotically stable critical point in the linear approximation, $f^o$ is moreover continuously differentiable with respect to $y$ in $D$ and has a domain of attraction $D^o \subset D$. If $x_o \in D^o$ we have

$$x(t) - y(t) = O(\delta(\epsilon)) \ , \ 0 \leqslant t < \infty$$

with $\delta(\epsilon) = o(1)$.

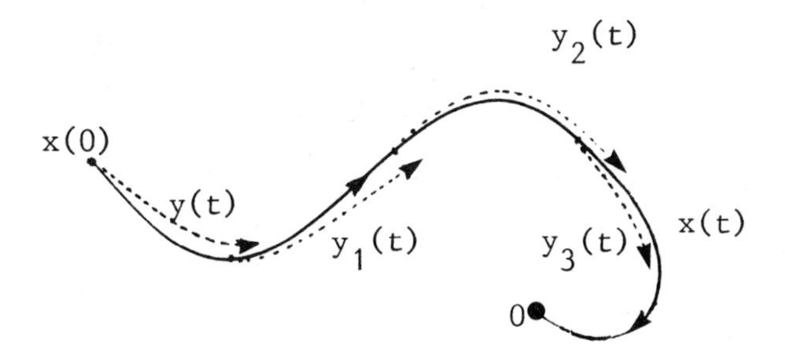


**Figure 4.2-1** *Solution x starts in* $x(0)$ *and attracts towards* 0.

**Proof**

Theorem 3.3.3 produces

$$\|x(t)-y(t)\| \leq \delta(\epsilon) \quad , \quad 0 \leq \epsilon t \leq L,$$

with $\delta(\epsilon)=o(1)$, the constant $L$ is independent of $\epsilon$ (note that $\delta$ is the order function as has been employed in § 3.3, multiplied by some constant independent of $\epsilon$). Putting $\tau=\epsilon t$ , $\frac{dy}{d\tau} = f^o(y)$ we know from lemma 1.4.6 that from a certain time $\tau=T$ on, the flow is exponentially contracting; T does not depend on $\epsilon$. Now we introduce the following partition of the time ( $=t$ )-axis

$$[0,\frac{T}{\epsilon}] \cup [\frac{T}{\epsilon},\frac{2T}{\epsilon}] \cup \cdots \cup [\frac{nT}{\epsilon},\frac{(n+1)T}{\epsilon}] \cup \cdots \quad , \; n=1,2, \cdots .$$

On each segment $I_n=[\frac{nT}{\epsilon},\frac{(n+1)T}{\epsilon}]$ we define $y_n$ as the solution of $\dot{y}=\epsilon f^o(y)$ with initial value $y_n(\frac{nT}{\epsilon})=x(\frac{nT}{\epsilon})$. For all finite $n$ we have from the averaging theorem

$$\|x(t)-y_n(t)\| \leq \delta(\epsilon) \quad , \quad t \in I_n. \qquad 4.2\text{-}1$$

If $\epsilon$ is small enough Lemma 1.4.6 produces on the other hand

$$\|y(t)-y_n(t)\|_{I_n} \leq \qquad 4.2\text{-}2$$

$$k\|y(\frac{nT}{\epsilon})-y_n(\frac{nT}{\epsilon})\| \leq k\|y(t)-y_n(t)\|_{I_{n-1}},$$

with $n=1,2, \cdots$ and $0<k<1$ and where $y_n$ has been continued on $I_{n-1}$ (the existence properties of the solutions permit this). The triangle inequality yields with 4.2-1 and 4.2-2

$$\|x(t)-y(t)\|_{I_n} \leq \delta + k\|y(t)-y_n(t)\|_{I_{n-1}}.$$

Using the triangle inequality again we find

$$\|x(t)-y(t)\|_{I_n} \leq \delta(\epsilon)+k\|x(t)-y(t)\|_{I_{n-1}} + k\|x(t)-y_n(t)\|_{I_{n-1}}$$

$$\leq (1+k)\delta(\epsilon)+k\|x(t)-y(t)\|_{I_{n-1}}.$$

To obtain the last estimate we used 4.2-1. We use this recursion relation to find

$$\|x(t)-y(t)\|_{I_n} \leq (1+k)\delta(\epsilon)(1+k+k^2+ \cdots +k^n).$$

Taking the limit for $n \to \infty$ finally yields that for $t \to \infty$

$$\|x(t)-y(t)\| \leq \frac{(1+k)}{1-k}\delta(\epsilon)$$

which completes the proof. □

Note that $\delta(\epsilon)$ in the estimate is asymptotically the order function as it arises in the averaging theorem. So in the periodic case we have an $O(\epsilon)$

estimate for $t \in [0,\infty)$. This applies for instance to the second example of § 1 if one transforms to polar coordinates. Somewhat more general: consider the autonomous system

$$\ddot{x} + x = \epsilon f(x,\dot{x}).$$

The averaging process has been carried out in example 2.7.1. Putting $x = r\cos(t+\psi)$ , $\dot{x} = -r\sin(t+\psi)$ we found after averaging the equations

$$\frac{d\tilde{r}}{dt} = -\epsilon f_1(\tilde{r}) \ , \quad \frac{d\tilde{\psi}}{dt} = -\epsilon \frac{f_2(\tilde{r})}{\tilde{r}}.$$

A critical point $\tilde{r} = r_o$ of the first equation will never be asymptotically stable in the linear approximation as $\tilde{\psi}$ has vanished from the equations. However, introducing polar coordinates $x = r\cos\theta$ , $\dot{x} = -r\sin\theta$ we find the first order equation

$$\frac{d\tilde{r}}{d\theta} = -\epsilon f_1(\tilde{r}).$$

If the critical point $\tilde{r} = r_o$ is asymptotically stable in the linear approximation we can apply our theorem. E.g. for the *van der Pol* equation (§ 2.7) we find

$$\frac{d\tilde{r}}{d\theta} = \tfrac{1}{2}\epsilon\tilde{r}(1 - \tfrac{1}{4}\tilde{r}^2).$$

There are two critical points: 0 and 2. The origin is unstable but $\tilde{r} = 2$ (corresponding with the limit cycle) has eigenvalue $-\epsilon$. Our theorem applies and we have

$$r(\theta) - \tilde{r}(\theta) = O(\epsilon) \ , \quad \theta \in [\theta_o, \infty)$$

for the orbits starting in $D^o$ in the domain of attraction.

## 4.3. An attractor in the original equation

Suppose that we started out with an equation of the form

$$\dot{x} = f(t,x) + \epsilon g(t,x)$$

and that the unperturbed equation ($\epsilon = 0$) contains an attracting critical point while satisfying the conditions of the *Poincaré-Lyapunov* theorem (Chapter 1). This case is even easier to handle than the case of averaging with attraction as the *Poincaré-Lyapunov* domain of the attractor is reached on a time-scale of order 1. In our formulation we shift again the critical point to the origin.

### 4.3.1. Theorem

Consider the equation

$$\dot{x} = f(t,x) + \epsilon g(t,x) \ , \quad x(0) = x_o,$$

with $x, x_o \in \mathbf{R}^n$ ; $y = 0$ is an asymptotically stable solution in the linear approximation of the unperturbed equation

$$\dot{y} = f(t,y) = (A + B(t))y + h(t,y),$$

with $A$ a constant $n \times n$ -matrix with all eigenvalues having negative real part, $B(t)$ is a continuous $n \times n$-matrix with the property

$$\lim_{t \to \infty} \|B(t)\| = 0.$$

$D$ is the domain of attraction of $x=0$. The vectorfield $h$ is continuous with respect to $t$ and $x$ and continuously differentiable with respect to x in $\mathbf{R}^+ \times D$, while

$$h(t,x) = o(\|x\|) \quad as \quad \|x\| \to 0, \; uniformly \; in \; t;$$

$g(t,x)$ is continuous in $t$ and $x$ and Lipschitz- continuous with respect to $x$ in $\mathbf{R}^+ \times D$. Choosing the initial value $x_o$ in the interior part of $D$ and adding to the unperturbed equation $y(0)=x_o$ we have

$$x(t) - y(t) = O(\epsilon) \quad , \; t \geqslant 0.$$

**Proof**

The solution $y(t)$ will be contained in the *Poincaré-Lyapunov* domain around $y=0$ for $t \geqslant T$. Note that $T$ does not depend on $\epsilon$ as the unperturbed equation does not depend on $\epsilon$. We use the partition of the time-axis

$$[0,T] \bigcup [T,2T] \bigcup \cdots \bigcup [nT,(n+1)T] \bigcup \cdots.$$

According to Lemma 2 of § 2.2 we have

$$x(t) - y(t) = O(\epsilon) \quad , \quad 0 \leqslant t \leqslant T.$$

From this point on we use exactly the same reasoning as formulated in Theorem 4.2.1. □

This is a first order result but of course, if the right hand side of the equation is sufficiently smooth, we can improve the accuracy by straightforward expansions. So here a naive use of perturbation techniques yields a uniformly valid result.

**4.3.2. Example (Cf. Example 2.4.1)**

Suppose two species are living in a region with a restricted supply of food. A slight interaction takes place as the first species occasionally predates on the second one. A simple model is given by

$$\dot{x} = ax - x^2 + \epsilon xy \qquad x, x(0) > 0,$$

$$\dot{y} = by - y^2 - \epsilon xy \qquad y, y(0) > 0.$$

The unperturbed equations are

$$\dot{x}_o = ax_o - x_o^2,$$

$$\dot{y}_o = by_o - y_o^2,$$

with asymptotically stable critical point $x_o = a, y_o = b$. The conditions of theorem 2 have been satisfied; expanding $x(t) = \sum_{n=0}^{\infty} \epsilon^n x_n(t)$ , $y(t) = \sum_{n=0}^{\infty} \epsilon^n y_n(t)$ we find

$$x(t) - \sum_{n=0}^{N} \epsilon^n x_n(t) = O(\epsilon^{N+1}) \quad , \quad t \geqslant 0,$$

$$y(t) - \sum_{n=0}^{N} \epsilon^n y_n(t) = O(\epsilon^{N+1}) \quad , \quad t \geqslant 0.$$

It is easy to compute $x_o(t), y_o(t)$ ; the higher order terms are obtained as the solutions of linear equations.

## 4.4. Contracting maps

We shall now formulate the results of § 2 in terms of mappings instead of vectorfields. This framework enables us to recover Theorem 4.2.1; Moreover one can use this idea to obtain new results. Consider again a differential equation of the form

$$\dot{x} = \epsilon f(t(,x) \qquad x \in D \subset \mathbf{R}^n.$$

Supposing that $f$ is a *KBM*-vectorfield we have the averaged equation

$$\dot{y} = \epsilon f^o(y) \qquad y \in D^o \subset D.$$

Again $f^o(y)$ has an attracting critical point, say $y=0$ and we know from Lemma 1.4.6 that under certain conditions there is a neighborhood $\Omega$ of $y=0$ where the phase-flow is actually contracting exponentially. This provides us with a contracting map of $\Omega$ into itself. Indicating a solution $y$ starting at $t=0$ in $y_o$ by $y(t;y_o)$ we have the map

$$F_o(y_o) = y(t_1;y_o) \quad , y_o \in \Omega \ , \ t_1 > 0.$$

Here we have solutions $y$ which approximate $x(t;x_o)$ for $0 \leqslant \epsilon t \leqslant L$ if $x_o$ and $y_o$ are close enough. So we take $t_1 = \dfrac{L}{\epsilon}$ and we define

$$F_o(y_o) = y(\frac{L}{\epsilon};y_o).$$

In the same way we define the map $F_\epsilon$ by

$$F_\epsilon(x_o) = x(\frac{L}{\epsilon};x_o).$$

If $x_o - y_o = o(1)$ as $\epsilon \rightarrow 0$ we have clearly

$$\|F_o(y_o) - F_\epsilon(x_o)\| \leqslant C\delta_1(\epsilon) \quad \textit{with } \delta_1(\epsilon) = o(1).$$

We shall prove that for a contracting map $F_o$, repeated application of the maps $F_o$ and $F_\epsilon$ does not enlarge the distance between the iterates significantly. We define the iterates by the relations

$$F_\epsilon^1(x) = F_\epsilon(x),$$
$$F_\epsilon^{m+1} = F_\epsilon(F_\epsilon^m(x)) \ , \ m=1,2, \cdots .$$

This will provide us with a theorem for contracting maps, analogous to the *Eckhaus / Sanchez – Palencia* theorem that we proved in § 2.

An application might be as follows. The equation for $y$ written down above, is simpler than the equation for $x$. Still it may be necessary to take recourse to numerical integration to solve the equation for $y$. If the numerical integration scheme involves an estimate of the error on intervals of the time with length $\frac{L}{\epsilon}$, we may envisage the numerical procedure as providing us with another map $F_h$ which approximates the map $F_o$. Using the same technique as formulated in the proof of the theorem, one can actually show that the numerical approximations in this case are valid on $[0,\infty)$ and therefore also approximate the solutions of the original equation on the same interval. In the context of a study of successive substitutions for perturbed operator equations *van der Sluis* (Slu70a) developed ideas which are related to the results discussed here. We shall split the proof into several lemma's to keep the various steps easy to follow.

**4.4.1. Lemma**

Consider a family of maps $F_\epsilon : D \rightarrow \mathbf{R}^n$ , $\epsilon \in [0,\epsilon_o]$ with the following properties:

a) For all $x \in D$ we have

$$\|F_o(x) - F_\epsilon(x)\| \leqslant \delta(\epsilon)$$

with $\delta(\epsilon)$ an order function, $\delta = o(1)$ as $\epsilon \downarrow 0$;

b) There exist constants $k$ and $\mu$ , $0 \leqslant k < 1$ , $\mu \geqslant 0$ such that for all $x,y \in D$

$$\|F_o(x) - F_o(y)\| \leqslant k\|x - y\| + \mu$$

(Contraction-attraction property of the unperturbed flow);

c) There exists an interior domain $D^o \subset D$ , invariant under $F_o$ such that the distance between the boundaries of $D^o$ and $D$ exceeds

$$\frac{\mu + \delta(\epsilon)}{1 - k};$$

then, if $\|x - y\| \leqslant \frac{\mu + \delta(\epsilon)}{1 - k}$ and $x \in D, y \in D^o$, we have for $m \in \mathbf{N}$

$$\|F_\epsilon^m(x) - F_o^m(y)\| \leqslant \frac{\mu + \delta(\epsilon)}{1 - k}$$

**Proof**

We use induction. If $m=0$ the statement is true; assuming that the statement is true for $m$ we prove it for $m+1$.

$$\|F_\epsilon^{m+1}(x) - F_o^{m+1}(y)\| \leqslant$$

(using the triangle inequality)

$$\|F_\epsilon(F_\epsilon^m(x)) - F_o(F_\epsilon^m(x))\| + \|F_o(F_\epsilon^m(x)) - F_o(F_o^m(y))\|.$$

As $y \in D^o$ and $D^o$ is invariant under $F_o$ we have $F_o^m(y) \in D^o$ . It follows from assumption (c) and the induction hypothesis that $F_\epsilon^m(x) \in D$. So we can use (a) and (b) to obtain from the inequality above

$$\|F_\epsilon^{m+1}(x) - F_o^{m+1}(y)\| \leqslant \delta(\epsilon) + k\|F_\epsilon^m(x) - F_o^m(y)\| + \mu$$

$$\leqslant \delta(\epsilon) + k\,\frac{\mu + \delta(\epsilon)}{1-k} + \mu = \frac{\mu + \delta(\epsilon)}{1-k}.$$

□

**4.4.2. Lemma**

Consider the equation

$$\dot{x} = f^o(x) \qquad x(0) = \xi$$

and suppose that for a given constant $L \in \mathbf{R}$ we are able to find an $\epsilon$-approximation to $x(t+L)$, given $x(t)$. Assume that for given $L$, we can make this approximation as good as we want (e.g. because it depends on a small parameter at our choice). Let $x_o$ be an attractor, with the properties given in Lemma 4.4.1 (The map is constructed from the flow in the proof of this Lemma) and let $W(x_o)$ be its domain of attraction. Then it is possible to find a uniformly valid approximation of $x(t)$, $t \in [0, \infty)$ of $O(\epsilon)$ if $\xi$ is well inside $W(x_o)$, using our approximation scheme. The nature of the approximation is not important. One might think of numerical computation or asymptotic expansions.

**Proof**

Let $t_1$ be such that $x(t_1)$ is inside the *Poincaré-Lyapunov* neighborhood of $x_o$. We can approximate $x(t_1)$ with $O(\epsilon)$ -accuracy. Let $t_1$ be the new origin in time. Take $t_2$ such that the map $F_o$, induced by flowing out the equation

$$\dot{x} = f^o(x) \qquad x(0) = \xi$$

(i.e. if $x(t,\xi)$ is the solution, $F_o(\xi) = x(t_2, \xi)$) leaves the *Poincaré-Lyapunov* domain $D^o$ invariant and is contracting, with contraction factor $\kappa < 1$ ; The existence of such a $t_2$ follows from 1.4.4 We have

$$\|F_o(x) - F_o(y)\| \leqslant \kappa \|x - y\|, \quad x,y \in D^o.$$

Let $F_\epsilon$ denote the approximation process (given $\xi$, we approximate $x(t_2,\xi)$ and call the approximation $F_\epsilon(\xi)$). We can now apply Lemma 4.4.1, to approximate with $O(\frac{\epsilon}{1-\kappa})$ accuracy $x(mt_2,\xi)$ , $m\in\mathbf{N}$. Since $t_2$ is $O_S(1)$, the same estimate holds for $x(t,\xi)$ for all $t\in[0,\infty)$ using the $O(\epsilon)$-approximation procedure between the points $mt_2$ and $(m+1)t_2$ for arbitrary $m\in\mathbf{N}$.

## 4.5. Attracting limit-cycles

In this section we shall discuss problems where the averaged equation has a limit-cycle. It turns out that the theory for this case is like the case with the averaged system having a stable stationary point, except that it is not possible to approximate the angular variable (or the flow on the limit-cycle) uniformly on $[0,\infty)$; the approximation, however, is possible on intervals of length $\epsilon^{-N}$, with $N$ arbitrary (but, of course, $\epsilon$-independent). We shall only sketch the results without giving proofs. For technical details the reader is referred to (San80a).

We consider systems of the form

$$\dot{\phi} = \Omega(x) + \epsilon\Omega^{(1)}(\phi,x) \qquad \phi\in\mathbf{T}^m,$$

$$\dot{x} = \epsilon X(\phi,x) \qquad x\in D\subset\mathbf{R}^n$$

and as an example, illustrative for the theory, we shall take the *van der Pol*-equation

$$\ddot{x} + \epsilon(x^2-1)\dot{x} + x = 0, \qquad x(0)=x_o, \dot{x}(0)=v_o, x_o^2+v_o^2\neq 0.$$

Introducing polar coordinates

$$x = r\sin\phi,$$

$$\dot{x} = r\cos\phi,$$

we obtain

$$\dot{\phi}=1+\epsilon[-\frac{1}{2}\sin 2\phi+\frac{1}{8}r^2(2\sin 2\phi-\sin 4\phi)],$$

$$\dot{r}=\epsilon\frac{r}{2}[1+\cos 2\phi-\frac{r^2}{4}(1-\cos 4\phi)].$$

The 'second order averaged' equation of this vectorfield is (see § 3.5)

$$\dot{\phi}=1-\frac{\epsilon^2}{8}(\frac{11}{32}r^4-\frac{3}{2}r^2+1)+O(\epsilon^3),$$

$$\dot{r}=\epsilon\frac{r}{2}(1-\frac{r^2}{4})+O(\epsilon^3).$$

Neglecting the $O(\epsilon^3)$ term, the equation for $r$ represents a subsystem with attractor $r=2$ . The fact that the $O(\epsilon^3)$ term depends on another variable as well (i.e. on $\phi$), is not going to bother us in our estimates since $\phi(t)$ is bounded (the circle is compact). This means that on solving the equation

$$\dot{\tilde{r}} = \epsilon\frac{\tilde{r}}{2}(1-\frac{\tilde{r}^2}{4}), \qquad \tilde{r}(0)=r_o=(x_o^2+v_o^2)^{1/2}$$

this is going to give us an $O(\epsilon)$ -approximation to the $r$-component of the original solution, valid on $[0,\infty)$. (The fact that the $\epsilon^2$-terms in the $r$-equation vanish, does in no way influence the results). Using this approximation, we can obtain an $O(\epsilon)$-approximation for the $\phi$-component on $0 \leq \epsilon^2 t \leq L$ by solving

$$\dot{\tilde{\phi}} = 1 - \frac{\epsilon^2}{8}\left(\frac{11}{32}\tilde{r}^4 - \frac{3}{2}\tilde{r}^2 + 1\right) \quad , \quad \tilde{\phi}(0) = \phi_o = \arctan\left(\frac{x_o}{v_o}\right).$$

Although this equation is easy to solve the treatment can even be more simplified by noting that the attraction in the $r$-direction takes place on a time-scale $\frac{1}{\epsilon}$, while the slow fluctuation of $\phi$ occurs on a time scale $\frac{1}{\epsilon^2}$. This has as a consequence that to obtain an $O(\epsilon)$ approximation $\tilde{\phi}$ for $\phi$ on the time-scale $\frac{1}{\epsilon^2}$ we may take $\tilde{r}=2$ on computing $\tilde{\phi}$. To prove this one uses an exponential estimate on $|r(t)-2|$ and *Gronwall's* inequality. Thus we are left with the following simple system

$$\dot{\hat{\phi}} = 1 - \frac{\epsilon^2}{16}, \qquad \hat{\phi}(0) = \phi_o,$$

$$\dot{\tilde{r}} = \epsilon\frac{\tilde{r}}{2}\left(1 - \frac{\tilde{r}^2}{4}\right), \qquad \tilde{r}(0) = r_o.$$

For the general solution of the *van der Pol*-equation with $r_o > 0$ we find

$$x(t) = \frac{r_o e^{\frac{1}{2}\epsilon t}}{[1 + \frac{1}{4} r_o^2 (e^{\epsilon t} - 1)]^{\frac{1}{2}}} \cos\left(t - \frac{\epsilon^2}{16}t + \phi_o\right) + O(\epsilon)$$

on $0 \leq \epsilon^2 t \leq L$ . There is no obstruction against carrying out the averaging process to any higher order to obtain approximations valid on longer time-scales, to be expressed in inverse powers of $\epsilon$.

## 4.6. Additional examples

To illustrate the theory in this chapter we shall discuss here examples which exhibit some of the difficulties.

### 4.6.1. Example: perturbation of the linear terms.

We have excluded in our theory and examples the possibility of perturbing the linear part of the differential equation in such a way that the stability characteristics of the attractor change. This is an important point as is shown by the following adaptation of an example in *Robinson* (Rob81a).

Consider the linear system with constant coefficients

$$\dot{x} = A(\epsilon)x + B(\epsilon)x,$$

with

$$A(\epsilon) = \begin{bmatrix} -\epsilon^2 & \epsilon \\ 0 & -\epsilon^2 \end{bmatrix} \quad , \quad B(\epsilon) = \epsilon^3 \begin{bmatrix} 0 & 0 \\ a^2 & 0 \end{bmatrix},$$

where $a$ is a positive constant. Omitting the $O(\epsilon^3)$ term we find negative eigenvalues $(-\epsilon^2)$ so that in this 'approximation' we have attraction towards the trivial solution. For the full equation we find eigenvalues $\lambda_{1,2} = -\epsilon^2 \pm a\epsilon^2$. So if $0 < a < 1$ we have attraction and $x = 0$ is asymptotically stable; if $a > 1$ the trivial solution is unstable. In both cases the flow is characterized by a time-scale $\frac{1}{\epsilon^2}$.

### 4.6.2. Example: Duffing's equation with damping and forcing.

Consider the equation

$$\ddot{x} + \epsilon a\dot{x} + x - \epsilon x^3 = \epsilon h\cos((1+\epsilon\Delta)t),$$

with $a, h$ positive constants. Transforming in the usual way $x = r\cos((1+\epsilon\Delta)t + \psi), \dot{x} = -r\sin((1+\epsilon\Delta)t + \psi)$ we find

$$\dot{r} = -\epsilon\sin((1+\epsilon\Delta)t+\psi)[ar\sin((1+\epsilon\Delta)t+\psi) + r^3\cos^3((1+\epsilon\Delta)t+\psi) + h\cos((1+\epsilon\Delta)t)],$$

$$\dot{\psi} = -\frac{\epsilon}{r}\cos((1+\epsilon\Delta)t+\psi)[ar\sin((1+\epsilon\Delta)t+\psi) + r^3\cos^3((1+\epsilon\Delta)t+\psi) + h\cos((1+\epsilon\Delta)t)].$$

Averaging produces

$$\dot{\tilde{r}} = -\epsilon\frac{a}{2}\tilde{r} - \epsilon\frac{h}{2}\sin\tilde{\psi},$$

$$\dot{\tilde{\psi}} = -\epsilon\Delta - \epsilon\frac{3}{8}\tilde{r}^2 - \epsilon\frac{h}{2\tilde{r}}\cos\tilde{\psi}.$$

Stationary solutions correspond with critical points $P$ given by the equations

$$a\tilde{r} = -h\sin\tilde{\psi} \quad , \quad \Delta + \frac{3}{8}\tilde{r}^2 = -\frac{h}{2\tilde{r}}\cos\tilde{\psi}.$$

A stationary solution of the averaged equations corresponds with an approximation of a periodic solution of the original equation. Linearization of the vectorfield at $P$ leads to the matrix

$$\begin{pmatrix} -\frac{a}{2} & -\frac{h}{2}\cos\tilde{\psi} \\ -\frac{3}{4}\tilde{r} + \frac{h}{2\tilde{r}^2}\cos\tilde{\psi} & \frac{h}{2\tilde{r}}\sin\tilde{\psi} \end{pmatrix}_P = \begin{pmatrix} -\frac{a}{2} & (\Delta + \frac{3}{8}\tilde{r}^2)\tilde{r} \\ -\frac{9}{8}\tilde{r} - \frac{\Delta}{\tilde{r}} & -\frac{a}{2} \end{pmatrix}_P .$$

The eigenvalues of the matrix are

$$\lambda_\pm = -\frac{a}{2} \pm (-(\Delta + \frac{3}{8}\tilde{r}^2)(\Delta + \frac{9}{8}\tilde{r}^2))^{1/2}.$$

If we assume $\Delta$ to be zero, there is only one stationary point, and, as the eigenvalues have a negative real part, the corresponding periodic solution is asymptotically stable. The solution of the averaged equations corresponding with the periodic solution constitutes an $O(\epsilon)$-approximation valid for all time; the same holds for solutions starting in an interior part of the domain of attraction of $P$ .

### 4.6.3. Example: damping on various time-scales.

Consider the equation of an oscillator with a linear and a nonlinear damping

$$\ddot{x}+\epsilon^n a\dot{x}+\epsilon\dot{x}^3+x=0.$$

The importance of the linear damping is determined by the choice of $n$ ; $a$ is a positive constant. We consider various cases.

#### 4.6.3.1. $n=0$

Putting $\epsilon=0$ we have the equation $\ddot{y}+a\dot{y}+y=0$ . Applying Theorem 4.3.1 we have that if $y(0)=x(0),\dot{y}(0)=\dot{x}(0),y(t)$ represents an $O(\epsilon)$ approximation of $x(t)$ uniformly valid in time. A naive expansion

$$x(t)=y(t)+\epsilon x_1(t)+\epsilon^2\cdots$$

produces higher order approximations with uniform validity.

If $n>0$ we put $x=r\cos(t+\psi),\dot{x}=-r\sin(t+\psi)$ to find

$$\dot{r}=-\epsilon^n ar\sin^2(t+\psi)-\epsilon r^3\sin^4(t+\psi),$$

$$\dot{\psi}=-\epsilon^n a\sin(t+\psi)\cos(t+\psi)-\epsilon r^2\sin^3(t+\psi)\cos(t+\psi).$$

Note that $\dot{r}(t)\leqslant 0$.

#### 4.6.3.2. $n=1$

The terms on the right hand side are of the same order in $\epsilon$; averaging produces

$$\dot{\tilde{r}}=-\epsilon\frac{a}{2}\tilde{r}-\epsilon\frac{3}{8}\tilde{r}^3,$$

$$\dot{\tilde{\psi}}=0.$$

The solutions can be used as approximations valid on the time-scale $\frac{1}{\epsilon}$.

However, we have attraction in the $r$-direction and we can proceed in the spirit of the results in the preceding section. Introducing polar coordinates by $\phi=t+\psi$ we find that we can apply Theorem 4.2.1 to the equation for the orbits $(\frac{dr}{d\phi}=\cdots)$ . So $\tilde{r}$ represents a uniformly valid approximation of $r$ ; higher order approximations can be used to obtain approximations for $\psi$ or $\phi$ which are valid on a longer time-scale than $\frac{1}{\epsilon}$.

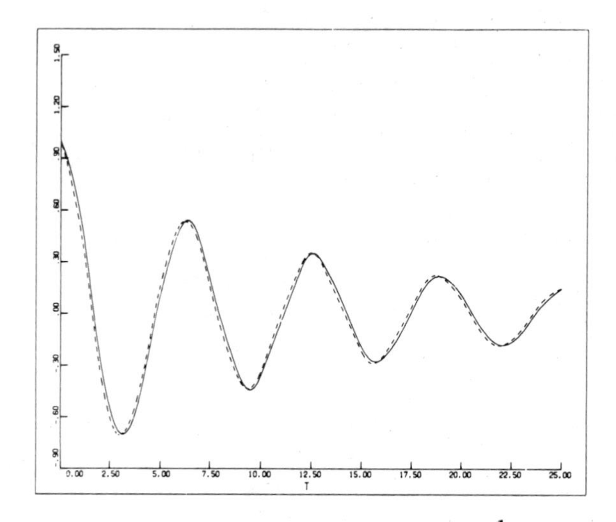


**Figure 4.6-1** *Linear attraction on the time scale* $\frac{1}{\epsilon^2}$ *for the equation* $\ddot{x}+x+\epsilon\dot{x}^3+3\epsilon^2\dot{x}=0$; $\epsilon=0.2$, $x(0)=1$, $\dot{x}(0)=0$. *The solution obtained by numerical integration has been drawn full line. The asymptotic approximation is indicated by - - - - and has been obtained from the equation, averaged to second order.*

**4.6.3.3.** $n=2$

The difficulty is that we cannot apply the preceding theorems at this stage as we have no attraction in the linear approximation. The idea is to use the linear higher order damping term nevertheless. Since the damping takes place on the time-scale $\frac{1}{\epsilon^2}$ the contraction constant $\kappa$ looks like $k=e^{\mu\epsilon}$ and therefore $\frac{1}{1-\kappa}=O_s(\frac{1}{\epsilon})$. We lose an order of magnitude in $\epsilon$ in our estimate, but we can win an order of magnitude by looking at the higher order approximation, which we are doing anyway, since we consider $O(\epsilon^2)$ terms.

The amplitude component of the averaged equation is

$$\dot{\rho}=-\frac{3}{8}\epsilon\rho^3-\frac{1}{2}\epsilon^2 a\rho\ ,\ \rho(0)=r_o$$

and we see that

$$\|r(t)-\rho(t)\|\leqslant C(L)\epsilon^2 \ \textit{on}\ 0\leqslant\epsilon t\leqslant L,$$

where $r$ is the solution of the nontruncated averaged equation. Using the contraction argument we find that

$$\|r(t)-\rho(t)\|\leqslant K(L)\epsilon \ \textit{for all}\ t\in[0,\infty),$$

where $K(L)=\frac{4C(L)}{aL}$. Since $r$ is an $O(\epsilon)$-approximation to the amplitude of the solution of the original equation for all time, we have solved our problem, at least for the amplitude.

# 5. Averaging over Spatial Variables: Systems with Slowly Varying Frequency and Passage through Resonance

## 5.1. Introduction

### 5.1.1. Examples

In our analysis of slowly varying systems we have developed up till now a theory for equations in the standard form

$$\dot{x} = \epsilon f(t,x).$$

In § 3.8 we studied an oscillator with slowly varying coefficients which could be brought into standard form after a rather special transformation of the time-scale. Systems with slowly varying coefficients, in particular varying frequencies, arise often in applications and we have to develop a systematic theory for these problems. Systems with slowly varying frequencies have been studied by *Mitropolsky* (Mit65a); an interesting example of passage through resonance has been considered by *Kevorkian* (Kev74a) using two-time scaling. Our treatment of the asymptotic estimates in this chapter is based on *Sanders* ((San78a) and (San79a) ) and forms an extension of the averaging theory of the periodic case as treated in chapters 2 and 3. We start by discussing a number of examples to see what the difficulties are. In §§ 2-4 we discuss the regular case which is relatively simple. In §§ 5-7 we introduce the reader to resonance problems.

**5.1.1.1. Example (*Einstein* pendulum)**

We consider a linear oscillator with slowly varying frequency

$$\ddot{x}+\omega^2(\epsilon t)x=0.$$

We put $\dot{x}=\omega y$. Differentiation produces $\ddot{x}=\dot{\omega}y+\omega\dot{y}$ and using the equation we find

$$\dot{y}=-\omega x-\frac{\dot{\omega}}{\omega}y.$$

We transform $(x,y)\mapsto(r,\phi)$ by

$$x=r\sin(\phi),$$

$$y=r\cos(\phi),$$

to find

$$\dot{r}=-\frac{\dot{\omega}}{\omega}r\cos^2(\phi),$$

$$\dot{\phi}=\omega+\frac{\dot{\omega}}{\omega}\sin(\phi)\cos(\phi).$$

Introducing $\tau=\epsilon t$ we have the third order system:

$$\dot{r}=-\frac{\epsilon}{\omega}\frac{d\omega}{d\tau}r\cos^2(\phi),$$

$$\dot{\tau}=\epsilon,$$

$$\dot{\phi}=\omega+\frac{\epsilon}{\omega}\frac{d\omega}{d\tau}\sin(\phi)\cos(\phi).$$

**5.1.1.2. Remark**

This system is of the form

$$\dot{x}=\epsilon X(\phi,x),\qquad x\in D\subset\mathbf{R}^n, \qquad 5.1.1\text{-}1$$

$$\dot{\phi}=\Omega(x)+\epsilon\cdots,\qquad \phi\in S^1,$$

with $x=(r,\tau)$; $\phi$ is an angular variable which is defined on the circle $S^1$; one could generalize the system to the case of $m$ angles with $\phi$ defined in the torus $T^m$.

**5.1.1.3. Remark**

One can remove the $0(\epsilon)$ terms in the equation for $\phi$ by a slightly different coordinate transformation. The price for this is an increase of the dimension of the system. Transform

$$x=r\sin(\phi+\psi),$$

$$y=r\cos(\phi+\psi),$$

with $\dot{\phi}=\omega$; we find

$$\dot{r} = -\frac{\epsilon}{\omega}\frac{d\omega}{d\tau} r\cos^2(\phi+\psi),$$

$$\dot{\psi} = \frac{\epsilon}{\omega}\frac{d\omega}{d\tau} \sin(\phi+\psi)\cos(\phi+\psi),$$

$$\dot{\tau} = \epsilon,$$

$$\dot{\phi} = \omega.$$

This form of the perturbation equations has some advantages in treating the resonance problems of §§ 5-7. For the sake of simplicity the theorems in this chapter concern system 5.1.1-1 with $\dot{\phi}=\Omega(x)$. In chapter 6 we treat system 5.1.1-1 with $\dot{\phi}=\Omega(x)+\epsilon\cdots$.

### 5.1.1.4. Remark

As $\phi\in S^1$ it seems natural to average the equation for $x$ in system 5.1.1-1 over $\phi$ to obtain an approximation of $x(t)$. It turns out that under certain conditions this procedure can be justified as we shall see later on.

### 5.1.1.5. Example, a nonlinear oscillator

It is a simple exercise to formulate in the same way the case of a nonlinear equation with a frequency governed by an independent equation:

$$\ddot{x}+\omega^2 x = \epsilon f(x,\dot{x},\epsilon t),$$

$$\dot{\omega} = \epsilon g(x,\dot{x},\epsilon t).$$

Put again $\dot{x}=\omega y$; by differentiation and using the equations we have

$$\dot{y} = -\omega x + \epsilon\frac{f}{\omega} - \epsilon y\frac{g}{\omega}.$$

Transforming

$$x = r\sin(\phi),$$

$$y = r\cos(\phi),$$

we find with $\tau=\epsilon t$ the fourth-order system

$$\dot{r} = \frac{\epsilon}{\omega}\cos(\phi)[f(r\cos(\phi),\omega r\cos(\phi),\tau) - r\cos(\phi)g(r\sin(\phi),\omega r\cos(\phi),\tau)],$$

$$\dot{\omega} = \epsilon g(r\sin(\phi),\omega r\cos(\phi),\tau),$$

$$\dot{\tau} = \epsilon,$$

$$\dot{\phi} = \omega - \frac{\epsilon}{\omega r}\sin(\phi)[f(r\cos(\phi),\omega r\cos(\phi),\tau) - r\cos(\phi)g(r\sin(\phi),\omega r\cos(\phi),\tau)].$$

Comparing with system 5.1.1-1 we have $n=3$, $x=(r,\omega,\tau)$.

We discuss now a problem in which two angles have to be used.

### 5.1.1.6. Example, an oscillator attached to a fly-wheel

The equations for such an oscillator have been discussed by *Goloskokow* and *Filippow* ((Gol71a), chapter 8.3).

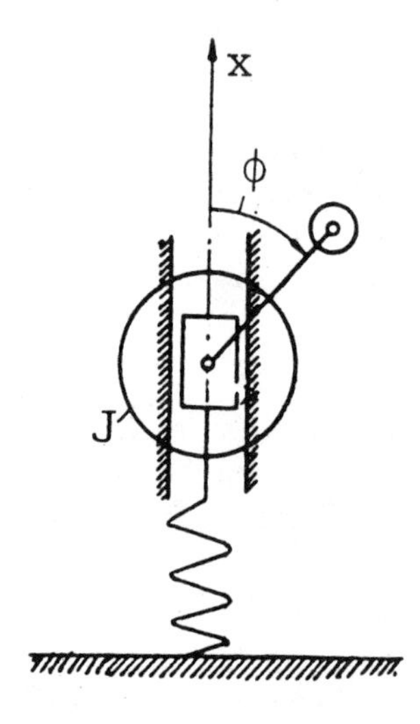


**Figure 5.1.1-1**

The frequency $\omega_o$ of the oscillator is a constant in this case; we assume that the friction, the nonlinear restoring force of the oscillator and several other forces are small. The equations of motion are

$$\ddot{x}+\omega_o^2 x=\epsilon F(\phi,\dot{\phi},\ddot{\phi},x,\dot{x},\ddot{x}),$$

$$\ddot{\phi}=\epsilon G(\phi,\dot{\phi},\ddot{\phi},x,\dot{x},\ddot{x}),$$

where

$$F=\frac{1}{m}[-f(x)-\beta\dot{x}+q_1(\dot{\phi}^2\cos(\phi)+\ddot{\phi}\sin(\phi)],$$

$$G=\frac{1}{J_o}[M(\dot{\phi})-M_w(\dot{\phi})]+q_2\sin(\phi)(\ddot{x}+g).$$

Here $\beta$, $q_1$ and $q_2$ are constants, $g$ is the gravitational constant, $J_o$ is the moment of inertia of the rotor. $M(\dot{\phi})$ represents the known static characteristic of the motor, $M_w(\dot{\phi})$ stands for the damping of rotational motion. The equations of motion can be written as

$$\dot{x}=\omega_o y,$$

$$\dot{y}=-\omega_o x+\frac{\epsilon}{\omega_o}F(\phi,\Omega,\dot{\Omega},x,\omega_o y,\omega_o\dot{y}),$$

$$\dot{\phi}=\Omega,$$

$$\dot{\Omega}=\epsilon G(\phi,\Omega,\dot{\Omega},x,\omega_o y,\omega_o\dot{y}).$$

We put $\phi=\phi_1$. As in the preceding examples we can put $x=r\sin(\phi_2)$,

$y = r\cos(\phi_2)$ to obtain

$$\dot{r} = \frac{\epsilon}{\omega_o}\cos(\phi_2)F(\phi_1,\Omega,\dot{\Omega},r\sin(\phi_2),\omega_o r\cos(\phi_2),\omega_o\dot{y}),$$

$$\dot{\Omega} = \epsilon G(\phi_1,\Omega,\dot{\Omega},r\sin(\phi_2),\omega_o r\cos(\phi_2),\omega_o\dot{y}),$$

$$\dot{\phi}_1 = \Omega,$$

$$\dot{\phi}_2 = \omega_o - \frac{\epsilon}{\omega_o r}\sin(\phi_2)F(\phi_1,\Omega,\dot{\Omega},r\sin(\phi_2),\omega_o r\cos(\phi_2),\omega_o\dot{y}).$$

$\dot{\Omega}$ and $\omega_o\dot{y}$ still have to be replaced using the equations of motion after which we can expand with respect to $\epsilon$. The system is of the form 5.1.1-1 with higher order terms added: $x = (r,\Omega)$, $\phi = (\phi_1,\phi_2)$. Again, it can be useful to simplify the equation for the angle $\phi_2$. We achieve this by starting with the equations of motion and putting

$$\phi = \phi_1 \quad , \quad x = r\sin(\phi_2 + \psi),$$

$$\phi_2 = \omega_o t \quad , \quad y = r\cos(\phi_2 + \psi).$$

The reader may verify that we obtain the fifth-order system

$$\dot{r} = \frac{\epsilon}{\omega_o}\cos(\phi_2+\psi)F(\phi_1,\Omega,\dot{\Omega},r\sin(\phi_2+\psi),\omega_o r\cos(\phi_2+\psi),\omega_o\dot{y}),$$

$$\dot{\psi} = -\frac{\epsilon}{\omega_o r}\sin(\phi_2+\psi)F(\phi_1,\Omega,\dot{\Omega},r\sin(\phi_2+\psi),\omega_o r\cos(\phi_2+\psi),\omega_o\dot{y}),$$

$$\dot{\Omega} = \epsilon G(\phi_1,\Omega,\dot{\Omega},r\sin(\phi_2+\psi),\omega_o r\cos(\phi_2+\psi),\omega_o\dot{y}),$$

$$\dot{\phi}_1 = \Omega,$$

$$\dot{\phi}_2 = \omega_2,$$

where $\dot{\Omega}$ and $\omega_o\dot{y}$ still have to be replaced using the equations of motion. We return to this example in § 8.

#### 5.1.1.7. Remark

Equations in the standard form $\dot{x} = \epsilon f(t,x)$, periodic in $t$, can be put in the form of system 5.1.1-1 in a trivial way. The equation is equivalent with

$$\dot{x} = \epsilon f(\phi,x),$$

$$\dot{\phi} = 1,$$

with the spatial variable $\phi \in S^1$.

### 5.1.2. Formal Theory

To derive a theory of asymptotic approximations for systems like 5.1.1-1 is more difficult than for systems in the standard form as treated in chapter 2 and 3. To see what the difficulties are, we start with a formal presentation. We put

$$x = y + \epsilon u(\phi, y),$$

where $u$ is to be an averaging transformation. So we have with 5.1.1-1

$$\dot{y} + \epsilon \frac{du}{dt} = \epsilon X(\phi, y + \epsilon u)$$

or

$$\dot{y} + \epsilon \frac{\partial u}{\partial \phi} \Omega(y + \epsilon u) + \epsilon \frac{\partial u}{\partial y} \dot{y} = \epsilon X(\phi, y + \epsilon u).$$

Expansion with respect to $\epsilon$ yields

$$(I + \epsilon \frac{\partial u}{\partial y}) \dot{y} = \epsilon X(\phi, y) - \epsilon \frac{\partial u}{\partial \phi} \Omega(y) + \epsilon^2 \cdots .$$

In the spirit of averaging it is natural to define

$$u(\phi, y) = \frac{1}{\Omega(y)} \int^{\phi} (X(\theta, y) - X^o(y)) d\theta, \qquad 5.1.2\text{-}1$$

where $X^o$ is the 'ordinary' average of $X$ over $\phi$, i.e. $X^o(y) = \int_{S^1} X(\phi, y) d\phi$. The equation becomes

$$\dot{y} = \epsilon X^o(y) + \epsilon^2 \cdots$$

and we add

$$\dot{\phi} = \Omega + \epsilon \cdots .$$

#### 5.1.2.1. Remark

Before analyzing these equations we note that one of the motivations for this formulation is that it is easy to generalize this formal procedure to multi-frequency systems. Assume $\phi = (\phi_1, \cdots, \phi_m)$ and let $X(\phi, x)$ be written as

$$X(\phi, x) = \sum_{i=1}^{m} X_i(\phi_i, x).$$

The equation for $\phi$ in 5.1.1-1 consists now of $m$ scalar equations of the form

$$\dot{\phi}_i = \Omega_i(x).$$

In the transformation $x = y + \epsilon u$ we put

$$u(\phi, y) = \sum_{i=1}^{m} u_i(\phi_i, y),$$

with

$$u_i(\phi_i, y) = \frac{1}{\Omega_i(y)} \int^{\phi_i} (X_i(\theta_i, y) - X_i^o(y)) d\theta_i .$$

The equation for $y$ then becomes

$$\dot{y}=\epsilon\sum_{i=1}^{m}X_i^o(y)+\epsilon^2\cdots. \qquad 5.1.2\text{-}2$$

One can obtain a formal approximation of the solutions of equation 5.1.2-2 by omitting the higher order terms and integrating the system

$$\dot{z}=\epsilon X^o(z)\ ,\ \dot{\psi}=\Omega(z).$$

To obtain in this way an asymptotic approximation of the solution of equation 5.1.1-1 we have to show that $u$ is bounded. Then however, we have to know a priori that $\Omega$ is bounded away from zero (Cf. equation 5.1.2-1). The following simple example illustrates the difficulty.

**5.1.2.2. Example (*Arnol'd* (Arn65a))**

Consider the scalar equations

$$\dot{x}=\epsilon(1-2\cos(\phi))\ ,\ x(0)=x_o\ ;\ x\in\mathbf{R},$$

$$\dot{\phi}=x \qquad ,\ \phi(0)=\phi_o\ ;\ \phi\in S^1$$

(written as a second order equation for $\phi$ the system becomes the familiar looking equation $\ddot{\phi}+2\epsilon\cos(\phi)=\epsilon$). The averaged equation 5.1.2-2 takes the form

$$\dot{y}=\epsilon+\epsilon^2\cdots,$$

$$\dot{\phi}=y+\epsilon\cdots.$$

So we would like to approximate $(\phi,x)$ by

$$(\phi_o+x_ot+\tfrac{1}{2}\epsilon t^2,x_o+\epsilon t).$$

The original equation has stationary solutions $(\frac{\pi}{3},0)$ and $(\frac{5\pi}{3},0)$ so if we put for instance $(\phi_o,x_o)=(\frac{\pi}{3},0)$, the error grows like $(\frac{1}{2}\epsilon t^2,\epsilon t)$. Note that the averaged equations contain no singularities; it can be shown however that the higher order terms do.

In the following we shall discuss the approximate character of the formal solutions in the simple case that $\Omega$ is bounded away from zero; this will be called the regular case.

## 5.2. Systems with slowly varying frequency in the regular case; the *Einstein* pendulum

We formulate and prove the following lemma which provides a useful perturbation scheme for system 5.1.1-1.

**5.2.1. Lemma**

Consider the equation with $C^1$-right hand sides

$$\dot{\phi}=\Omega(x) \quad , \phi(0)=\phi_o \ , \phi\in S^1, \tag{5.2-1}$$

$$\dot{x}=\epsilon X(\phi,x) \ , \ x(0)=x_o \ , \ x\in D\subset\mathbf{R}^n.$$

Suppose $0<m\leqslant \inf_{x\in D}|\Omega(x)|\leqslant \sup_{x\in D}|\Omega(x)|\leqslant M<\infty$ where $m$ and $M$ are $\epsilon$-independent constants. We transform

$$\phi(t)=\psi(t), \tag{5.2-2}$$

$$x(t)=y(t)+\epsilon u(\psi(t),y(t)),$$

with $(\psi,y)$ the solution of

$$\dot{\psi}=\Omega(y)+\epsilon R_1(\psi,y;\epsilon), \tag{5.2-3}$$

$$\dot{y}=\epsilon X^o(y)+\epsilon^2 R_2(\psi,y;\epsilon).$$

$R_1$ and $R_2$ are to be constructed later on and $\psi(0),y(0)$ are determined by 5.2-2. One defines

$$X^o(y)=\int_{S^1} X(\phi,y)d\phi$$

and

$$u(\psi,y)=\frac{1}{\Omega(y)}\int^{\psi}(X(\phi,y)-X^o(y))d\phi. \tag{5.2-4}$$

We use the convention $\int_{S^1} d\phi=1$. We choose the integration constant such that

$$\int_{S^1} u(\psi,y)d\psi=0.$$

Then $u,R_1$ and $R_2$ are uniformly bounded.

**Proof**

$u$ has been defined explicitly and is uniformly bounded because of the two-sided estimate for $\Omega$ and the integrand in 5.2-4 having zero average. $R_1$ and $R_2$ have been defined implicitly and will now be determined, at least to zeroth order in $\epsilon$. We differentiate the relations 5.2-2 and substitute the vectorfield 5.2-1.

$$\dot{\phi}=\Omega(x)$$

implies

$$\dot{\psi}=\Omega(y+\epsilon u(\psi,y))=\Omega(y)+\epsilon R_1(\psi,y;\epsilon),$$

with

$$\epsilon R_1=\Omega(y+\epsilon u(\psi,y))-\Omega(y).$$

For $\epsilon \downarrow 0$, $R_1$ becomes

$$R_1(\psi,y;0) = \nabla \Omega(y) u(\psi,y).$$

With the implicit function theorem we establish the existence and uniform boundedness of $R_1$ for $\epsilon \in (0,\epsilon_o]$.

For the $x$-component we have the following relations:

$$\dot{x} = \epsilon X(\phi,x) = \epsilon X(\psi,y+\epsilon u(\psi,y))$$

and

$$\begin{aligned}
\dot{x} &= \dot{y} + \epsilon \frac{\partial u}{\partial \psi} \dot{\psi} + \epsilon \nabla u \dot{y} \\
&= \epsilon X^o(y) + \epsilon^2 R_2(\psi,y;\epsilon) + \epsilon \frac{\partial u}{\partial \psi} (\Omega(y) + \epsilon R_1(\psi,y;\epsilon)) \\
&\quad + \epsilon \nabla u(\epsilon X^o(y)) + \epsilon^2 R_2(\psi,y;\epsilon)) \\
&= \epsilon X^o(y) + \epsilon^2 R_2(\psi,y;\epsilon) + \frac{\epsilon}{\Omega(y)}(X(\psi,y) - X^o(y))(\Omega(y) + \epsilon R_1(\psi,y;\epsilon)) \\
&\quad + \epsilon \nabla u(\epsilon X^o(y) + \epsilon^2 R_2(\psi,y;\epsilon)) \\
&= \epsilon X^o(y) + \epsilon^2 R_2(\psi,y;\epsilon) + \epsilon X(\psi,y) - \epsilon X^o(y) \\
&\quad + \frac{\epsilon^2}{\Omega(y)} R_1(\psi,y;\epsilon)(X(\psi,y;\epsilon) - X^o(y)) + \epsilon \nabla u(\epsilon X^o(y) + \epsilon^2 R_2(\psi,y;\epsilon)) \\
&= \epsilon X(\psi,y) + \epsilon^2 R_2(\psi,y;\epsilon) + \epsilon^2 R_1(\psi,y;\epsilon) \frac{\partial u}{\partial \psi}(\psi,y) \\
&\quad + \epsilon \nabla u(\epsilon X^o(y) + \epsilon^2 R_2(\psi,y;\epsilon)).
\end{aligned}$$

This gives the equation

$$\epsilon(I + \epsilon \nabla u) R_2 = X(\psi,y+\epsilon u(\psi,y)) - X(\psi,y) - \epsilon R_1 \frac{\partial u}{\partial \psi} - \epsilon \nabla u X^o.$$

In the limit $\epsilon \downarrow 0$, we can solve this:

$$\begin{aligned}
R_2(\psi,y;0) &= \nabla X u - R_1(\psi,y;0) \frac{\partial u}{\partial \psi} - \nabla u X^o \\
&= \nabla X u \; - \; \nabla \Omega u \frac{\partial u}{\partial \psi} \; - \; \nabla u X^o.
\end{aligned}$$

Using again the implicit function theorem we find the existence and uniform boundedness of $R_2$.

Transformation 5.2-2 has produced equation 5.2-3; later on, in chapter 6, we shall call this calculation a normalization process. We truncate equation 5.2-3 and we shall first prove the validity of the solution of the resulting equation as an asymptotic approximation to the solution of the nontruncated equation 5.2-3.

**5.2.2. Lemma**

Consider equation 5.2-3 from Lemma 5.2.1 with the same conditions and solution $(\psi, y)$. Let $(\zeta, z)$ be the solution of

$$\dot{\zeta}=\Omega(z), \qquad \zeta(0)=\zeta_o,$$

$$\dot{z}=\epsilon X^o(z), \qquad z(0)=z_o\ ,\ z\in D.$$

Remark that the initial values of both systems need not be the same. Then

$$\|y-z\|\leqslant(\|y_o-z_o\|+\epsilon^2 t\|R_2\|)e^{\epsilon Lt},$$

where

$$\|R_2\|=\sup_{(\psi,y,\epsilon)\in S^1\times D\times(0,\epsilon_o]}|R_2(\psi,y;\epsilon)|$$

and $L$ is the *Lipschitz*-constant of $X^o$. If $z_o=y_o+O(\epsilon)$ this implies

$$y(t)=z(t)+O(\epsilon) \text{ on the time-scale } \frac{1}{\epsilon}.$$

Furthermore

$$|\psi-\zeta|\leqslant|\psi_o-\zeta_o|$$

$$+\|\nabla\Omega\|(\|y_o-z_o\|t+\frac{\epsilon}{L}t\|R_1\|e^{\epsilon Lt}-\frac{1}{L^2}t\|R_2\|e^{\epsilon Lt}+\frac{1}{L^2})+\epsilon t\|R_1\|.$$

On the time-scale $\frac{1}{\epsilon}$, this only implies an $O(1)$-estimate for the angular variable.

**Proof**

The proof is standard. We write

$$y(t)-z(t)=y_o+\epsilon\int_0^t X^o(y(\tau))d\tau$$

$$+\epsilon^2\int_0^t R_2(\psi(\tau),y(\tau);\epsilon)d\tau-z_o-\epsilon\int_0^t X^o(z(\tau))d\tau$$

or

$$\|y(t)-z(t)\|\leqslant\|y_o-z_o\|+\epsilon\int_0^t\|X^o(y(\tau))-X^o(z(\tau))\|d\tau+\epsilon^2\|R_2\|t.$$

Noting that $\|X^o(y)-X^o(z)\|\leqslant L\|y-z\|$ and applying *Gronwall*'s Lemma 1.3.1 produces the desired result. The same type of reasoning produces the estimate for $\psi-\zeta$.□

From a combination of the two Lemmas we obtain an averaging Theorem:

**5.2.3. Theorem**

Consider the equations with initial values

$$\dot{\phi}=\Omega(x), \qquad \phi(0)=\phi_o \ , \ \phi\in S^1, \qquad 5.2\text{-}5$$

$$\dot{x}=\epsilon X(\phi,x), \quad x(0)=x_o \ , \ x\in D\subset \mathbf{R}^n.$$

Suppose

$$0<m\leqslant \inf_{x\in D}|\Omega(x)|\leqslant \sup_{x\in D}|\Omega(x)|\leqslant M<\infty,$$

where $m$ and $M$ are $\epsilon$-independent constants. Let $(\zeta,z)$ be the solution of

$$\dot{\zeta}=\Omega(z), \qquad \zeta(0)=\zeta_o,$$

$$\dot{z}=\epsilon X^o(z), \quad z(0)=z_o \ , \ z\in D,$$

where

$$X^o(z)=\int_{S^1} X(\phi,z)d\phi.$$

Then, if $z(t)$ remains in $D^o\subset D$

$$x(t)=z(t)+O(\epsilon) \textit{ on the time-scale } \frac{1}{\epsilon}.$$

Furthermore $\phi(t)=\zeta(t)+O(\epsilon te^{\epsilon t})$.

**Proof**

Transform equations 5.2-5 with Lemma 5.2.1, $(\phi,x)\mapsto(\psi,y)$ and apply Lemma 5.2.2.□

**5.2.4. Remark**

One can generalize Theorem 5.2.3 to include the case $\phi\in\mathbf{T}^m$ or to the nonperiodic case $\phi\in\mathbf{R}$, where some generalized average over $\phi$ exists. These extensions have been left to the reader.

**5.2.5. Example: the *Einstein* pendulum**

Consider the equation with slowly varying frequency

$$\ddot{x}+\omega^2(\epsilon t)x=0,$$

with initial conditions given; we assume as in Theorem 5.2.3

$$0<m\leqslant\omega(\epsilon t)\leqslant M<\infty.$$

In § 5.1.1 we found in this case the perturbation equations

$$\dot{r}=-\frac{\epsilon}{\omega}\frac{d\omega}{d\tau}r\cos^2(\phi),$$

$$\dot{\tau}=\epsilon,$$

$$\dot{\phi}=\omega+\frac{\epsilon}{\omega}\frac{d\omega}{d\tau}\sin(\phi)\cos(\phi).$$

Averaging over $\phi$ we find the equations

$$\dot{\tilde{r}} = -\frac{\epsilon}{2\omega}\frac{d\omega}{d\tau}\tilde{r},$$

$$\dot{\tilde{\tau}} = \epsilon.$$

After integration we find

$$\tilde{r}(t)\omega^{1/2}(\epsilon t) = r_o\omega^{1/2}(0)$$

and $r(t) = \tilde{r}(t) + O(\epsilon)$ on the time-scale $\frac{1}{\epsilon}$. In the original coordinates we may write

$$\omega(\epsilon t)x^2 + \frac{\dot{x}^2}{\omega(\epsilon t)} = \textit{constant} + O(\epsilon) \textit{ on the time-scale } \frac{1}{\epsilon},$$

which is a well-known adiabatic invariant of the system (the energy of the system changes linearly with the frequency). Note that in § 3.8 we have treated these problems using a special time-like variable; the advantage here is that there is no need to find such special transformations.

## 5.3. Higher order approximation in the regular case

The estimates obtained in the preceding Lemma's and in Theorem 5.2.3 can be improved. this is particularly useful in the case of the angle $\phi$ for which only an $O(1)$ estimate has been obtained on the time-scale $\frac{1}{\epsilon}$. First we have a second order version of Lemma 5.2.1:

### 5.3.1. Lemma

Consider the equation

$$\dot{\phi} = \Omega(x), \qquad \phi(0) = \phi_o\ ,\ \phi \in S^1, \qquad 5.3\text{-}1$$

$$\dot{x} = \epsilon X(\phi, x), \quad x(0) = x_o\ ,\ x \in D \subset \mathbf{R}^n$$

and assume the conditions of Lemma 5.2.1. for the solutions $\phi, x$ of equation 5.3-1 we can write

$$x(t) = y(t) + \epsilon u^{(1)}(\psi(t), y(t)) + \epsilon^2 u^{(2)}(\psi(t), y(t)), \qquad 5.3\text{-}2$$

$$\phi(t) = \psi(t) + \epsilon v^{(1)}(\psi(t), y(t)),$$

where $\psi$ and $y$ are solutions of

$$\dot{\psi} = \Omega(y) + \epsilon^2 R_1(\psi, y; \epsilon)\ ,\ \psi(0) = \psi_o, \qquad 5.3\text{-}3$$

$$\dot{y} = \epsilon X^{1o}(y) + \epsilon^2 X^{2o}(y) + \epsilon^3 R_2(\psi, y; \epsilon)\ ,\ y(0) = y_o.$$

$X^{2o}$ is defined by

$$X^{2o}(y) = \int_{S^1} (\nabla X u^{(1)} - X\frac{\nabla\Omega}{\Omega}u^{(1)})d\phi.$$

$u^{(1)}(\psi, y)$ is defined as $u$ in Lemma 5.2.1, equation 5.2-4, $X^{1o}(y)$ as $X^o$ in

Lemma 5.2.1.

**Proof**

We present the formal computation and we shall not give all the technical details as in the proof of Lemma 5.2.1. From equations 5.3-(1-2) we have

$$\dot{\phi}=\Omega(y+\epsilon u^{(1)}+\epsilon^2 u^{(2)})=\Omega(y)+\epsilon\nabla\Omega u^{(1)}+O(\epsilon^2).$$

On the other hand, differentiating the second part of transformation 5.3-2 yields

$$\dot{\phi}=\dot{\psi}+\epsilon\frac{\partial v^{(1)}}{\partial\psi}\dot{\psi}+\epsilon\frac{\partial v^{(1)}}{\partial y}\dot{y}$$

$$= \textit{(using 5.3-3)}\ \Omega(y)+\epsilon\frac{\partial v^{(1)}}{\partial\psi}\Omega+O(\epsilon^2).$$

Comparing the two expressions for $\dot{\phi}$ we define

$$v^{(1)}(\psi,y)=\frac{1}{\Omega(y)}\int^{\psi}\nabla\Omega(y)u^{(1)}(\phi,y)d\phi.$$

In the same way we have from equation 5.3-1 with transformation 5.3-2

$$\dot{x}=\epsilon X(\psi+\epsilon v^{(1)},y+\epsilon u^{(1)}+\epsilon^2 u^{(2)})$$

$$=\epsilon X(\psi,y)+\epsilon^2\frac{\partial X}{\partial\psi}v^{(1)}+\epsilon^2\nabla X u^{(1)}+O(\epsilon^3).$$

Differentiating the first part of transformation 5.3-2 yields

$$\dot{x}=\dot{y}+\epsilon\frac{\partial u^{(1)}}{\partial\psi}\dot{\psi}+\epsilon\nabla u^{(1)}\dot{y}+\epsilon^2\frac{\partial u^{(2)}}{\partial\psi}\dot{\psi}+\epsilon^2\nabla u^{(2)}\dot{y}$$

and with equation 5.3-3

$$=\epsilon X^{1o}+\epsilon^2 X^{2o}+\epsilon\frac{\partial u^{(1)}}{\partial\psi}\Omega+\epsilon^2\nabla u^{(1)}X^{1o}+\epsilon^2\frac{\partial u^{(2)}}{\partial\psi}\Omega+O(\epsilon^3).$$

Comparing the two expressions for $\dot{x}$ we have indeed

$$u^{(1)}(\psi,y)=\frac{1}{\Omega(y)}\int^{\psi}(X(\phi,y)-X^{1o}(y))d\phi$$

and moreover we find

$$u^{(2)}=\frac{1}{\Omega}\int^{\psi}\left(\frac{\partial X}{\partial\phi}v^{(1)}+\nabla X u^{(1)}-\nabla u^{(1)}X^{1o}-X^{2o}\right)d\phi.$$

There is no need to compute $v^{(1)}$ explicitly at this stage; we find, requiring $u^{(2)}$ to have zero average

$$X^{2o}=\int_{S^1}\left(\frac{\partial X}{\partial\phi}v^{(1)}+\nabla X u^{(1)}-\nabla u^{(1)}X^{1o}\right)d\phi$$

$$=\int_{S^1}\left(-X\frac{\partial v^{(1)}}{\partial\phi}+\nabla X u^{(1)}\right)d\phi$$

$$= \int_{S^1} (\nabla X - X \frac{\nabla \Omega}{\Omega}) u^{(1)} d\phi.$$

□

Following the same reasoning as in § 2 we first approximate the solutions of equation 5.3-3.

**5.3.2. Lemma**

Consider equation 5.3-3 with initial values

$$\dot{\psi} = \Omega(y) + \epsilon^2 R_1(\psi, y; \epsilon), \qquad \psi(0) = \psi_o,$$
$$\dot{y} = \epsilon X^{1o}(y) + \epsilon^2 X^{2o}(y) + \epsilon^3 R_2(\psi, y; \epsilon)\ , \ y(0) = y_o.$$

Let $(\zeta, z)$ be the solution of the truncated system

$$\dot{\zeta} = \Omega(z)\ , \qquad \zeta(0) = \zeta_o,$$
$$\dot{z} = \epsilon X^{1o}(z) + \epsilon^2 X^{2o}(z)\ , \ z(0) = z_o$$

then

$$\|z(t) - y(t)\| \leqslant (\|z_o - y_o\| + \epsilon^3 t \|R_2\|) e^{\epsilon L t},$$

where $L$ is the *Lipschitz*-constant of $X^{1o} + \epsilon X^{2o}$. If $z_o = y_o + O(\epsilon^2)$, this implies that

$$y(t) = z(t) + O(\epsilon^2) \textit{ on the time-scale } \frac{1}{\epsilon}.$$

Furthermore

$$|\psi(t) - \zeta(t)| \leqslant |\psi_o - \zeta_o| + \|\nabla \Omega\| \|z(t) - y(t)\| t + \epsilon^2 t \|R_1\|.$$

If $z_o = y_o + O(\epsilon^2)$ and $\zeta_o = \psi_o + O(\epsilon)$ this produces an $O(\epsilon)$-estimate on the time-scale $\frac{1}{\epsilon}$ for the angular variable $\psi$.

**Proof**

The proof is standard and runs along precisely the same lines as the proof of Lemma 5.2.2. □

We are now able to approximate the solutions of the original equation 5.3-1 to a higher order precision.

**5.3.3. Theorem (second order averaging)**

Consider equation 5.3-1

$$\dot{\phi} = \Omega(x), \qquad \phi(0) = \phi_o\ , \ \phi \in S^1,$$
$$\dot{x} = \epsilon X(\phi, x), \quad x(0) = x_o\ , \ x \in D \subset \mathbf{R}^n$$

and assume the conditions of Lemma 5.3-1. Following Lemma 5.3.2 we define $(\zeta, z)$ as the solution of

$$\dot{\zeta}=\Omega(z) \qquad , \zeta(0)=\phi_o,$$
$$\dot{z}=\epsilon X^{1o}(z)+\epsilon^2 X^{2o}(z) \ , \ z(0)=x_o-\epsilon u^{(1)}(\phi_o,x_o).$$

Then, on the time-scale $\frac{1}{\epsilon}$,

$$\phi(t)=\zeta(t)+O(\epsilon),$$
$$x(t)=z(t)+\epsilon u^{(1)}(\zeta(t),z(t))+O(\epsilon^2).$$

**Proof**

If $(\psi,y)$ is defined as in Lemma 5.3.2, then

$$x_o=y_o+\epsilon u^{(1)}(\psi_o,y_o)+O(\epsilon^2),$$

so

$$z_o-y_o=x_o-\epsilon u^{(1)}(\phi_o,x_o)-y_o$$
$$=\epsilon u^{(1)}(\psi_o,y_o)-\epsilon u^{(1)}(\phi_o,x_o)=O(\epsilon^2),$$
$$\zeta_o-\psi_o=O(\epsilon).$$

Applying Lemma 5.3.2 we have on the time-scale $\frac{1}{\epsilon}$

$$\psi(t)=\zeta(t)+O(\epsilon),$$
$$y(t)=z(t)+O(\epsilon^2).$$

Since $x(t)=y(t)+\epsilon u^{(1)}(\psi(t),y(t))+O(\epsilon^2)$, we obtain the estimate of the theorem. Note that we also have

$$x(t)=z(t)+O(\epsilon)$$

on the time-scale $\frac{1}{\epsilon}$.□

## 5.4. Generalization of the regular case; an example from celestial mechanics

In a number of problems in mechanics one encounters equations in which $\Omega$ in equation 5.3.1 also depends on the angle $\phi$. For instance

$$\dot{\phi}=\Omega(\phi,x),$$
$$\dot{x}=\epsilon X(\phi,x).$$

We shall show here how to obtain a first-order approximation for $x(t)$. The right hand sides may also depend explicitly on $t$. The computations in that case become rather complicated and we shall not explore such problems here. Note however, that if the dependence on $t$ is periodic we can interpret $t$ as an angle $\theta$ while adding the equation $\dot{\theta}=1$. One might wonder, as in § 2, if $\Omega$ is bounded away from zero, why not divide the equation for $\dot{x}$ by $\dot{\phi}$ and simply average over $\phi$. The answer is first, that one would have

have then an approximation in the time-like variable $\phi$ as independent variable; the estimate would still have to be extended to the behavior in $t$. More importantly, it is not clear how by such a simple approach one can generalize the procedure to multi-frequency systems ($\phi=(\phi_1, \cdots, \phi_m)$) and to cases where the right hand side depends on $t$.

In the following we shall not repeat all the technical details of § 2 but we shall try to convey that the general ideas of § 2 apply in this case.

**5.4.1. Lemma**

Consider the equations with $C^1$-right hand sides

$$\dot{\phi}=\Omega(\phi,x)\ ,\ \phi(0)=\phi_o\ ,\ \phi\in S^1, \qquad 5.4\text{-}1$$

$$\dot{x}=\epsilon X(\phi,x)\ ,\ x(0)=x_o\ ,\ x\in D\subset\mathbf{R}^n.$$

Suppose $0<m\leqslant \inf_{x\in D}\inf_{\phi\in S^1}|\Omega(\phi,x)|\leqslant \sup_{x\in D}\sup_{\phi\in S^1}|\Omega(\phi,x)|\leqslant M<\infty$ where $m$ and $M$ are $\epsilon$-independent constants. Transform

$$\phi(t)=\psi(t),$$

$$x(t)=y(t)+\epsilon u(\psi(t),y(t)).$$

Let $(\psi,y)$ be the solution of

$$\dot{\psi}=\Omega(\psi,y)+\epsilon R_1(\psi,y;\epsilon), \qquad 5.4\text{-}2$$

$$\dot{y}=\epsilon X^o(y)+\epsilon^2 R_2(\psi,y;\epsilon).$$

Define

$$X^o(y)=\frac{\int_{S^1}\frac{X(\phi,y)}{\Omega(\phi,y)}d\phi}{\int_{S^1}\frac{1}{\Omega(\phi,y)}d\phi} \qquad 5.4\text{-}3$$

and

$$u(\psi,y)=\int^{\psi}\frac{1}{\Omega(\phi,y)}[X(\phi,y)-X^o(y)]d\phi, \qquad 5.4\text{-}4$$

where the integration constant is such that

$$\int_{S^1}u(\psi,y)d\psi=0;$$

then $u$, $R_1$ and $R_2$ are uniformly bounded.

**Proof**

The vectorfield $u$ has been defined explicitly by equation 5.4-4 and is uniformly bounded because of the two-sided estimate for $\Omega$ and the integrand in 5.4-4 having zero average. Differentiation of the relation between $x$ and $y$ and substitution of the vectorfield in 5.4-2 produces

$$\dot{y}+\epsilon\frac{\partial u}{\partial\phi}(\phi,y+\epsilon u)\Omega(\phi,y+\epsilon u)+\epsilon\nabla u(\phi,y+\epsilon u)\dot{y}=\epsilon X(\phi,y+\epsilon u).$$

Expanding with respect to $\epsilon$ and using 5.4-(3-4) yields

$$\dot{y}=\epsilon X^o(y)+\epsilon^2R_2(\psi,y;\epsilon).$$

In the same way

$$\dot{\psi}=\Omega(\psi,y+\epsilon u)=\Omega(\psi,y)+\epsilon R_1(\psi,y;\epsilon).$$

The existence and uniform boundedness of $R_1$ and $R_2$ are established as in Lemma 5.2.1. □

We now formulate an analogous version of Lemma 5.2.2 for equation 5.4-2.

**5.4.2. Lemma**

Consider equation 5.4-2 with initial conditions

$$\dot{\psi}=\Omega(\psi,y)+\epsilon R_1(\psi,y;\epsilon)\ ,\ \psi(0)=\psi_o\ ,\ \psi\in S^1,$$

$$\dot{y}=\epsilon X^o(y)+\epsilon^2R_2(\psi,y;\epsilon)\ ,\ y(0)=y_o\ ,\ y\in D\subset\mathbf{R}^n.$$

Let $(\zeta,z)$ be the solution of the truncated system

$$\dot{\zeta}=\Omega(\zeta,z),\qquad \zeta(0)=\zeta_o, \tag{5.4-5}$$

$$\dot{z}=\epsilon X^o(z),\quad z(0)=z_o.$$

Then

$$\|y(t)-z(t)\|\leqslant(\|y_o-z_o\|+\epsilon^2t\|R_2\|)e^{L\epsilon t},$$

where $L$ the *Lipschitz*-constant of $X^o$ is. If $z_o=y_o+O(\epsilon)$ this implies

$$y(t)=z(t)+O(\epsilon)\ \textit{on the time-scale}\ \frac{1}{\epsilon}.$$

If moreover $X^o=0$

$$\|y(t)-z(t)\|\leqslant\|y_o-z_o\|+\epsilon^2t\|R_2\|, \tag{5.4-6}$$

which implies the possibility of extension of the time-scale of validity.

**5.4.3. Trivial remark**

On estimating $|\psi(t)-\zeta(t)|$ one finds an $O(1)$ estimate on the time-scale 1, which result is even worse than the one in Lemma 5.2.2. However, though the error grows faster in this case, one should realize that both $\psi$ and $\zeta$ are in $S^1$ so that the error never exceeds $O(1)$.

**Proof**

The proof runs along the same lines as for Lemma 5.2.2. We note that if $X^o=0$ we put

$$y(t)-z(t)=y_o-z_o+\epsilon^2\int_0^t R_2(\psi(\tau,y(\tau);\epsilon)d\tau,$$

which directly produces 5.4-6. □

Apart from the expressions 5.4-(3-4) no new results have been obtained thus far. However we had to formulate Lemma 5.4.1-2 to obtain the following Theorem.

**5.4.4. Theorem**

Consider the equations with initial values

$$\dot{\phi}=\Omega(\phi,y)+\epsilon R_1(\phi,y;\epsilon)\ ,\ \phi(0)=\phi_o\ ,\ \phi\in S^1,$$

$$\dot{y}=\epsilon X^o(y)+\epsilon^2 R_2(\phi,y;\epsilon)\ ,\ y(0)=y_o\ ,\ y\in D\subset \mathbf{R}^n.$$

Suppose $0<m\leqslant \inf_{x\in D}\inf_{\phi\in S^1}|\Omega(\phi,x)|\leqslant \sup_{x\in D}\sup_{\phi\in S^1}|\Omega(\phi,x)|\leqslant M<\infty$ where $m$ and $M$ are $\epsilon$-independent constants. Let $(\zeta,z)$ be the solution of the truncated system

$$\dot{\zeta}=\Omega(\zeta,z),\quad \zeta(0)=\zeta_o,$$

$$\dot{z}=\epsilon X^o(z),\quad z(0)=z_o.$$

where

$$X^o(y)=\frac{\int_{S^1}\frac{X(\phi,y)}{\Omega(\phi,y)}d\phi}{\int_{S^1}\frac{1}{\Omega(\phi,y)}d\phi}.$$

Then if $z(t)$ remains in $D^o$

$$x(t)=z(t)+O(\epsilon)\ \textit{on the time-scale}\ \frac{1}{\epsilon}.$$

If $X^o=0$, $z(t)=z_o$ and we have, if we put $z_o=x_o$,

$$x(t)=x_o+O(\epsilon)+O(\epsilon^2 t).$$

**Proof**

Apply Lemma 5.4.1-2. □

To illustrate the preceding theory we shall discuss an example from celestial mechanics. The equations contain time $t$ explicitly but this causes no complications as this is in the form of slow time $\epsilon t$.

**5.4.5. Example (two-body problem with variable mass)**

Consider the *Newtonian* two-body problem in which the total mass $m$ decreases monotonically and slowly with time. If the loss of mass is isotropic and is removed instantly from the system, the equation of motion in polar coordinates $\theta,r$ is

$$\ddot{r}=-\frac{Gm}{r^2}+\frac{c^2}{r^3},$$

with angular momentum integral

$$r^2\dot{\theta}=c.$$

$G$ is the gravitational constant. To express the slow variation with time we put $m=m(\tau)$ with $\tau=\epsilon t$. *Hadjidemetriou* (Had63a) derived the perturbation equations for the orbital elements $e$ (eccentricity), $E$ (eccentric anomaly, a phase angle) and $\omega$ (angle indicating the direction of the line of apsides); an alternative derivation has been given by *Verhulst* (Ver75a). We have

$$\frac{de}{dt}=-\epsilon\frac{(1-e^2)\cos(E)}{1-e\cos(E)}\frac{1}{m}\frac{dm}{d\tau}, \qquad 5.4\text{-}7a$$

$$\frac{d\omega}{dt}=-\epsilon\frac{(1-e^2)\sin(E)}{e(1-e\cos(E))}\frac{1}{m}\frac{dm}{d\tau}, \qquad 5.4\text{-}7b$$

$$\frac{dE}{dt}=\frac{(1-e^2)^{\frac{3}{2}}}{1-e\cos(E)}\frac{G^2}{c^3}m^2+\frac{\sin(E)}{e(1-e\cos(E))}\frac{1}{m}\frac{dm}{d\tau}, \qquad 5.4\text{-}7c$$

$$\frac{d\tau}{dt}=\epsilon. \qquad 5.4\text{-}7d$$

$E$ plays the part of the angle $\phi$ in the standard system 5.4-1; note that here $n=3$. To apply the preceding theory the first term on the right hand side of equation 5.4-7c must be bounded away from zero. This means that for perturbed elliptic orbits we have the restriction $0<\alpha<e<\beta<1$ with $\alpha,\beta$ independent of $\epsilon$. Then we can apply Theorem 5.4.4 with $x=(e,\omega,\tau)$; in fact 5.4-7c is somewhat more complicated than the equation for $\phi$ in Theorem 5.4.4 but this does not affect the first-order computation. We find

$$\int_{S^1}\frac{X(\phi,x)}{\Omega(\phi,x)}d\phi=0$$

for the first two equations and $\dot{\tau}=\epsilon$ as it should be. So the eccentricity $e$ and the position of the line of apsides $\omega$ are constant with accuracy $O(\epsilon)$ on the time-scale $\frac{1}{\epsilon}$. In other words: we have proved that the quantities $e$ and $\omega$ are adiabatic invariants for these perturbed elliptic orbits if we exclude the nearly-circular and nearly-parabolic cases. To obtain non-trivial behavior one has to calculate higher order approximations in $\epsilon$ or first order approximations on a longer time-scale or one has to study the excluded domains in $e$: $[0,\alpha]$ and $[\beta,1]$. This work has been carried out and for further details we refer the reader to (Ver75a); see also appendix 8.7.

## 5.5. Introduction to resonance problems

### 5.5.1. Introduction

In the introduction to this chapter we met a difficulty while applying straightforward averaging techniques to the problem at hand. We then studied the case when this difficulty could not happen, i.e. $\Omega(x)$ does not vanish, calling this the regular case. We now return to the problem where $\Omega$

can have zeros or can be small. We cannot present a complete theory as such a theory is not available, so we rather aim at introducing the reader to the relevant concepts. This may serve as an introduction to the literature. To be more concrete, we will study the following equations.

$$\dot{\phi}=\Omega(x), \qquad \phi(0)=\phi_o\ ,\ \phi\in S^1,$$

$$\dot{x}=\epsilon X(\phi,x), \quad x(0)=x_o\ ,\ x\in D\subset\mathbf{R}^n.$$

If we try to average this equation in the sense of § 2, our averaging transformation becomes singular at the zeros of $\Omega$. If $\Omega$ is (near) zero, we say that the system is in resonance. This terminology derives from the fact that in many applications the angle $\phi$ is in reality the difference between two angles: to say that $\Omega\sim 0$ is equivalent to saying that the frequencies of the two angles are about equal, or that they are in resonance. We shall meet this point of view again in the chapter on *Hamiltonian* systems.

Well, this vanishing of $\Omega$ is certainly a problem, but, as is so often the case, it has a local character. And if the problem is local, we can use this information to simplify our equations by *Taylor*-expansion. To put it more formally, we define the *resonance manifold* $N$ as follows.

$$N=\{(\phi,x)\in S^1\times D\subset S^1\times\mathbf{R}^n \mid \Omega(x)=0\}.$$

$N$ is only a manifold in the very original sense of the word, that is of the solution set of an equation. $N$ is in general *not* invariant under the flow of the differential equation; in general this flow is transversal to $N$ in a sense which will be made clear in the sequel. To study the behavior of the equations locally we need local variables, a concept originating from boundary layer theory; for an introduction to these concepts see (Eck79a).

### 5.5.2. The inner expansion

To illustrate the technique in its simplest form we put $n=1$ and we assume that $\Omega(0)=0$. A local variable $\xi$ is obtained by scaling $x$ near 0.

$$x=\delta(\epsilon)\xi,$$

where $\delta(\epsilon)$ is an order function with $\lim_{\epsilon\downarrow 0}\delta(\epsilon)=0$; $\xi$ is supposed to describe the *inner region* or *boundary layer*. In the local variable the equations read

$$\dot{\phi}=\Omega(\delta(\epsilon)\xi)=\Omega(0)+\delta(\epsilon)\frac{d\Omega}{dx}(0)\xi+O(\delta^2)$$

$$=\delta(\epsilon)\frac{d\Omega}{dx}(0)\xi+O(\delta^2),$$

$$\delta(\epsilon)\dot{\xi}=\epsilon X(\phi,0)+O(\epsilon\delta).$$

Truncating these equations we obtain the *inner vectorfield* on the right hand sides:

$$\dot{\phi}=\delta(\epsilon)\frac{d\Omega}{dx}(0)\xi,$$

$$\dot{\xi}=\delta^{-1}(\epsilon)\epsilon X(\phi,0).$$

Solutions of this last set of equations are called *formal inner expansions* or *formal local expansions* (If the asymptotic validity has been shown we leave out the *formal).* The corresponding second order equation is

$$\ddot{\phi}=\epsilon\frac{d\Omega}{dx}(0)X(\phi,0)$$

so that the natural time-scale of the inner equation is $\frac{1}{\epsilon^{1/2}}$. A consistent choice for $\delta(\epsilon)$ might then be $\delta(\epsilon)=\epsilon^{1/2}$, but we shall see that there is some need to take the size of the boundary layer around $N$ somewhat larger.

### 5.5.2.1. Example

Consider the two-dimensional system

$$\Omega(x)=x,$$

$$X(\phi,x)=\alpha(x)-\beta(x)\sin(\phi).$$

Expanding near the resonance manifold $x=0$ we have the inner equation

$$\ddot{\phi}+\epsilon\beta(0)\sin(\phi)=\epsilon\alpha(0).$$

For $\beta(0)>0$ and $|\frac{\alpha(0)}{\beta(0)}|<1$ the phase flow is sketched in figure 5.2.2-1.

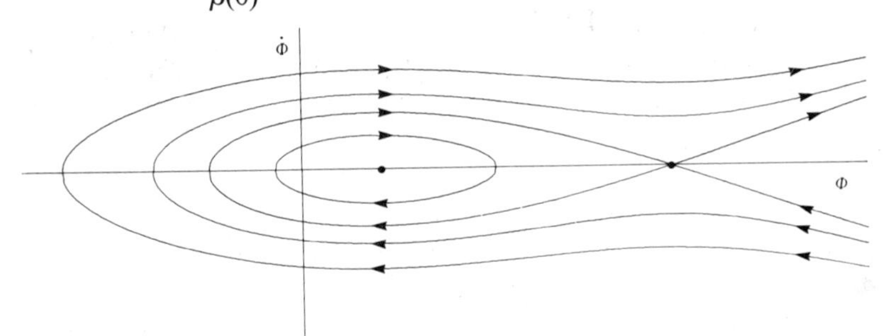


**Figure 5.5.2-1**

Note that $N$ corresponds with $\dot{\phi}=0$. If the solution enters the boundary layer near the stable manifold of the saddle point it might stay for a long while in the inner region, in fact much longer than the natural time-scale. In such cases we will be in trouble as the theory discussed so far does not extend beyond the *natural time-scales.*

### 5.5.3. The outer expansion

Away from the resonance manifold and its neighborhood where $\Omega(x)\sim 0$ we have the *outer region* and corresponding *outer expansion* of the solution. In fact we have already seen this outer expansion as it can be taken to be the solution of the averaged equation in the sense of § 2. The averaging process provides us with valid answers if one keeps fair distance from the resonance manifold. This explains the term outer, since one is always looking out from the singularity. We shall see however, that if we try to extend the validity of the outer expansion in the direction of the resonance manifold we meet the restriction

$$\frac{\epsilon}{\Omega^2(x)} = o(1).$$

This means that if $d(x,N)$ is the distance of $x$ to the resonance manifold $N$ we have

$$\frac{\epsilon}{d^2(x,N)} = o(1).$$

So, if we take $\delta(\epsilon)=\epsilon^{1/2}$, as suggested in the discussion of the inner expansion, we cannot extend the averaging results to the boundary of the inner domain. Thus we shall consider an inner region of size $\delta$, somewhat larger than $\epsilon^{1/2}$, but this poses the problem of how to extend the validity of the inner expansion to the time-scale $\frac{1}{\delta(\epsilon)}$ in the region $c_1\epsilon^{1/2} \leqslant d(x,N) \leqslant c_2\delta(\epsilon)$. We need this longer time-scale for the solution to have time to leave the boundary layer. This problem, which is by no means trivial, can be solved using the special structure of the inner equations.

### 5.5.4. The composite expansion

Once we obtain the inner and outer expansion we proceed to construct a *composite expansion.* To do this we add the inner expansion to the outer expansion while subtracting the common part, the so called *inner-outer expansion.* For the foundations of this process of composite expansions and matching we refer again to (Eck79a), chapter 3. In formula this becomes

$$x_C = x_I + x_O - x_{IO},$$

$x_C$ : *the composite expansion,*

$x_I$ : *the inner expansion,*

$x_O$ : *the outer expansion,*

$x_{IO}$ : *the inner-outer expansion.*

In the inner region, $x_C$ has to look like $x_I$, so $x_{IO}$ should look like $x_O$ to cancel the outer expansion; this means that $x_{IO}$ should be the inner expansion of the outer expansion, i.e. the outer expansion reexpanded in the inner variables. Analogous reasoning applies to the composite expansion in the outer region.

This type of expansion procedure can be carried out for vectorfields. We shall define the *inner-outer vectorfield* as the averaged inner expansion or, equivalently, the around $N$ expanded averaged equation. That the averaged equation can be expanded at all near $N$ may surprise us but it turns out to be possible at least to first order. The second order averaged equation might be singular at $N$. The solution of the inner-outer vectorfield is then the inner-outer expansion. From the validity of the averaging method and the expansion method, which we shall prove, follows the validity of the composite expansion method, that is we can write the original solution $x$ as

$$x = x_C + O(\eta(\epsilon)) \quad \eta = o(1).$$

If the solution enters the inner domain and leaves it at the other side, then we speak of *passage through resonance.* We can only describe this asymptotically if the inner vectorfield is transversal to $N$. Otherwise the asymptotic solution, i.e. the composite expansion, cannot pass through the resonance, even though the real solution might be able to do this (be it on a much longer time-scale than the natural time-scale of the inner expansion).

## 5.6. Remarks on higher-dimensional problems

### 5.6.1. Introduction

In our discussion thus far we have assumed $x$ and $\phi$ to be scalar variables. If the dimension of the spatial variable $n$ is larger than one, one should choose a minimal number of coordinates transversal to $N$ (measuring the distance to $N$) and split $\mathbf{R}^n$ accordingly. For instance if $n=2$ and

$$\dot{\phi} = x_1(1 + x_2^2),$$

the dimension of the system is three. $N$ is determined by $x_1 = 0$ and $x_1$ can be used as a transversal coordinate; $x_2$ plays no essential part. The remaining coordinates, like $x_2$ in this example, remain unscaled in the inner region and have variations of $O(\epsilon^{1/2})$ on the time-scale $\dfrac{1}{\epsilon^{1/2}}$. They play no part in the asymptotic analysis of the resonance to first order and we can concentrate on the lower dimensional problem, in this case with dimension 2.

### 5.6.2. The case of more than one angle

If $\phi \in T^m, m > 1$, the situation is complicated and the theory is far from complete. We outline the problems for the case $m=2$ and $n$ spatial variables:

$$\begin{aligned}
\dot{\phi}_1 &= \Omega_1(x), \\
\dot{\phi}_2 &= \Omega_2(x), \qquad\qquad \phi_1, \phi_2 \in S^1, \\
\dot{x} &= \epsilon X(\phi_1, \phi_2, x) \quad , \; x \in \mathbf{R}^n.
\end{aligned}$$

The right hand side of the equation for $x$ is expanded in a complex *Fourier* series; we have

$$\dot{x} = \epsilon \sum_{k,l=-\infty}^{\infty} c_{kl}(x) e^{i(k\phi_1 + l\phi_2)}.$$

Averaging over the angles *outside* the resonances

$$k\Omega_1 + l\Omega_2 = 0$$

leads to the averaged equation

$$\dot{y} = \epsilon c_{oo}(y).$$

A resonance arises for instance if

$$k\Omega_1(x) + l\Omega_2(x) = 0 \quad , \ k,l \in \mathbf{Z}. \qquad 5.6.2\text{-}1$$

This resonance condition has as a consequence that, in principle, an infinite number of resonance domains can be found. In each of these domains we have to localize around the resonance manifold given by 5.6.2-1. Locally we construct an expansion with respect to the resonant variable $k\phi_1 + l\phi_2 = \Phi$ while averaging over all the other combinations. Note that we have assumed that the resonance domains are disjunct. So, locally we have again a problem with one angle $\Phi$ and what remains is the problem of obtaining a global approximation to the solution.

Before treating some simple examples we mention a case which occurs quite often in practice. Suppose we have *one* angle $\phi$ and a perturbation which is also a periodic function of $t$, period 1.

$$\dot{\phi} = \Omega(x) + \cdots,$$

$$\dot{x} = \epsilon X(t,\phi,x).$$

It is natural to introduce now two angles
$\phi_1 = \phi$, $\phi_2 = t$ and adding the equation

$$\dot{\phi}_2 = 1.$$

The resonance condition 5.6.2-1 becomes in this case

$$k\Omega(x) + l = 0.$$

So each rational value assumed by $\Omega(x)$ corresponds with a resonance domain provided that the corresponding $k,l$-coefficient arises in the *Fourier* expansion of $X$.

We conclude this discussion with two simple examples given by *Arnol'd* (Arn65a); in both cases $n=2$, $m=2$.

### 5.6.2.1. Example of resonance locking

The equations are

$$\dot{\phi}_1 = x_1,$$

$$\dot{\phi}_2 = x_2,$$

$$\dot{x}_1 = \epsilon,$$

$$\dot{x}_2 = \epsilon\cos(\phi_1 - \phi_2).$$

The resonance condition 5.6.2-1 reduces to the case $k=1$, $l=-1$:

$$x_1 = x_2.$$

There are two cases to consider: First suppose we start in the resonance manifold, so $x_1(0)=x_2(0)$, and let $\phi_1(0)=\phi_2(0)$. Then the solutions are easily seen to be

$$\phi_1(t)=\phi_1(0)+x_1(0)t+\tfrac{1}{2}\epsilon t^2,$$

$$\phi_2(t)=\phi_1(t),$$

$$x_1(t)=x_2(t)=x_1(0)+\epsilon t.$$

The solutions are locked into resonance, due to the special choice of initial conditions. The second case arises when we start outside the resonance domain: $x_1(0)-x_2(0)=a\neq 0$ with $a$ independent of $\epsilon$. Averaging over $\phi_1-\phi_2$ produces the equations

$$\dot\psi_1=y_1,$$

$$\dot\psi_2=y_2,$$

$$\dot y_1=\epsilon,$$

$$\dot y_2=0,$$

with the solutions

$$\psi_1(t)=\phi_1(0)+x_1(0)t+\tfrac{1}{2}\epsilon t^2,$$

$$\psi_2(t)=\phi_2(0)+x_2(0)t,$$

$$y_1(t)=x_1(0)+\epsilon t,$$

$$y_2(t)=x_2(0).$$

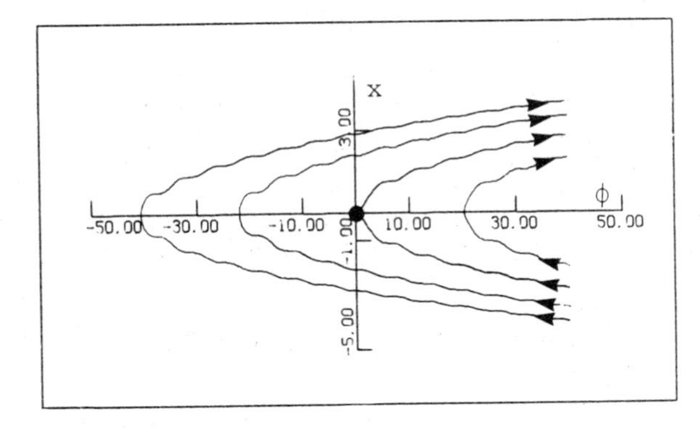


**Figure 5.6.2-1** *The x,ϕ-phase plane for example 5.6.2.1;* $\epsilon=0.1$. *Resonance locking takes place at* (0,0).

To establish the asymptotic character of these formal approximations, note first that

$$x_1(t)=y_1(t).$$

Furthermore we introduce $x=x_1-x_2$, $\phi=\phi_1-\phi_2$ to obtain

$$\dot{\phi}=x,$$

$$\dot{x}=\epsilon(1-\cos(\phi)),$$

which has the integral

$$x^2=a^2+2\epsilon\phi-2\epsilon\sin(\phi).$$

At the same time we have

$$(y_1-y_2)^2=a^2+2a\epsilon t+\epsilon^2t^2=a^2+2\epsilon(\psi_1-\psi_2).$$

This expression agrees with the integral to $O(\epsilon)$ for all time. Although the approximate integral constitutes a valid approximation of the integral which exists for the system, this is still not enough to characterize the individual orbits on the integral manifold. We omit the technical discussion for this detailed characterization.

#### 5.6.2.2. Example of forced passage through resonance

The equations are

$$\dot{\phi}_1=x_1+x_2,$$

$$\dot{\phi}_2=x_2,$$

$$\dot{x}_1=\epsilon,$$

$$\dot{x}_2=\epsilon\cos(\phi_1-\phi_2).$$

The resonance condition reduces to

$$x_1=0.$$

There are two cases to consider:

**The case $a=x_1(0)>0$**

1) Let $b=\phi_1(0)-\phi_2(0)$. Since $\dot{x}_1>0$ we have $x_1(t)>0$ for all $t>0$. Thus there will be no resonance. Using averaging, we can show that $x_2(t)=x_2(0)+O(\epsilon)$, but we can also see this from solving the original equations:

$$x_1(t)=x_1(0)+\epsilon t,$$

$$\phi_1(t)-\phi_2(t)=b+x_1(0)t+\tfrac{1}{2}\epsilon t^2,$$

$$x_2(t)=x_2(0)+\epsilon\int_0^t\cos(b+a\tau+\tfrac{1}{2}\epsilon\tau^2)d\tau=x_2(0)+O(\epsilon). \qquad 5.6.2\text{-}2$$

The last estimate is valid for all time and can be obtained by transforming $\sigma=b+a\tau+\tfrac{1}{2}\epsilon\tau^2$, followed by partial integration.

**5.6.2.3. Exercise**

Where exactly do we use the fact that $a>0$ ?

**The case** $a<0$.

2) Using formula 5.6.2-2 we can take the limit for $t\to\infty$ to compute the change in $x_2$ induced by the passage through the resonance. The integral is a well known *Fresnel*-integral and the result is

$$\lim_{t\to\infty} x_2(t)=x_2(0)+\left(\frac{\pi\epsilon}{2}\right)^{1/2}\cos\left(b-\frac{a^2}{2\epsilon}+\frac{\pi}{4}\right).$$

The $\epsilon^{1/2}$-contribution of the resonance agrees with the analysis given in § 5.5.1.

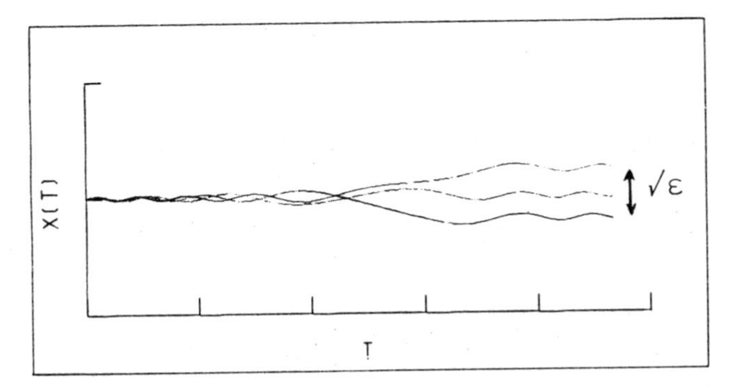


**Figure 5.6.2-2** *Solutions* $x=x_2(t)$ *based on equation 5.6.2-2 with* $b=0$, $\epsilon=.1$ *and* $O(\epsilon)$ *variations of a around* $-2$.

Remark that the dependency of the shift in $x_2$ is determined by the orbit, as it should be, since changing the initial point in time should have no influence on this matter.

## 5.7. Analysis of the inner and outer expansion; passage through resonance

After the intuitive reasoning in the preceding sections we shall now develop the asymptotic analysis of the expansions outside and in the resonance region. We shall also discuss the rather intricate problem of matching these expansions and thus the phenomenon of passage through resonance.

**5.7.1. Lemma**

Suppose that one can split $x$ in $(\eta,\xi)\in\mathbf{R}^{n-1}\times\mathbf{R}$ such that $\Omega(\eta,0)=0$. Denote the $\eta$-component of $X$ by $X_{/\!/}$ and the $\xi$ -component by $X_{\vdash}$. Then consider the equation

$$\dot{\phi}=\Omega(\eta,\xi), \tag{5.7-1}$$

$$\dot{\xi}=\epsilon X_{\vdash}(\phi,\eta,\xi),$$

$$\dot{\eta}=\epsilon X_{/\!/}(\phi,\eta,\xi)$$

in a $\epsilon^{1/2}$-neighborhood of $\xi=0$. Introduce the following norm $\|\ \|$:

$$\|((\phi,\eta),\xi)\| = \|(\phi,\eta)\| + \frac{1}{\epsilon^{1/2}}\|\xi\|.$$

Denote by $(\Delta\phi,\Delta\eta,\Delta\xi)$ the difference of the solution of equation 5.7-1 and the inner equation:

$$\dot{\phi}=\frac{\partial\Omega}{\partial\xi}(\eta,0)\xi,$$

$$\dot{\xi}=\epsilon X_{\vdash}(\phi,\eta,0), \qquad 5.7\text{-}2$$

$$\dot{\eta}=0.$$

Then we have the estimate

$$\|(\Delta\phi,\Delta\eta,\Delta\xi)(t)\| \leqslant (\|(\Delta\phi,\Delta\eta,\Delta\xi)(0)\| + R\epsilon t)e^{K\epsilon^{1/2}t}.$$

with $R$ and $K$ constants independent of $\epsilon$.

### 5.7.2. Remark

We do not assume any knowledge of the initial conditions here, since this is only part of a larger scheme; this estimate indicates however that considering approximations on the time-scale $\frac{1}{\epsilon^{1/2}}$ one has to know the initial conditions at least with accuracy $o(\epsilon^{1/2})$.

**Proof**

Let $(\phi,\eta,\xi)$ be the solution of the original equation 5.7-1 and $(\bar{\phi},\bar{\eta},\bar{\xi})$ the solution of the inner equation 5.7-2. Then we can estimate the difference using the *Gronwall* Lemma:

$$\|(\Delta\phi,\Delta\eta,\Delta\xi)(t)\| \leqslant \|(\Delta\phi,\Delta\eta,\Delta\xi)(0)\| + \int_0^t \{\|\Omega(\eta,\xi)-\frac{\partial\Omega}{\partial\xi}(\bar{\eta},0)\bar{\xi}\|$$

$$+\epsilon^{1/2}\|X_{\vdash}(\phi,\eta,\xi)-X_{\vdash}(\bar{\phi},\bar{\eta},0)\| + \epsilon\|X_{/\!/}(\phi,\eta,\xi)\|\}d\tau$$

$$\leqslant \|(\Delta\phi,\Delta\eta,\Delta\xi)(0)\| + \int_0^t \{\|\frac{\partial\Omega}{\partial\xi}\|\|\xi-\bar{\xi}\| + C\epsilon^{1/2}(\|\phi-\bar{\phi}\|+\|\eta-\bar{\eta}\|)+R\epsilon\}d\tau$$

$$\leqslant \|(\Delta\phi,\Delta\eta,\Delta\xi)(0)\| + K\epsilon^{1/2}\int_0^t \|(\Delta\phi,\Delta\eta,\Delta\xi)(\tau)\|\,d\tau + R\int_0^t \epsilon\, d\tau$$

and this implies

$$\|(\Delta\phi,\Delta\eta,\Delta\xi)(t)\| \leqslant (\|(\Delta\phi,\Delta\eta,\Delta\xi)(0)\| + R\epsilon t)e^{K\epsilon^{1/2}t}.$$

□

**5.7.3. Exercise**

Generalize this Lemma to the case where $\xi = O(\delta(\epsilon))$. Is it possible to get estimates on a larger time-scale than $\frac{1}{\epsilon^{1/2}}$ ?

In the next Lemma we shall generalize Lemma 5.3.1; the method of proof is the same, but we are more careful about the inverse powers of $\Omega$ appearing in the perturbation scheme.

**5.7.4. Lemma**

Consider the equation

$$\dot{\phi} = \Omega(x), \qquad \phi(0) = \phi_o \ , \ \phi \in S^1,$$
$$\dot{x} = \epsilon X(\phi, x), \quad x(0) = x_o \ , \ x \in D \subset \mathbf{R}^n.$$

Then we can write the solution of this equation $(\phi, x)$ as follows:

$$(\phi, x)(t) = (\psi(t) + \epsilon v^{(1)}(\psi(t), y(t)),$$
$$y(t) + \epsilon u^{(1)}(\psi(t), y(t)) + \epsilon^2 u^{(2)}(\psi(t), y(t))), \qquad 5.7\text{-}3$$

where $(\psi, y)$ is the solution of

$$\dot{\psi} = \Omega(y) + \epsilon^2 R_1(\psi, y; \epsilon), \qquad \psi(0) = \psi_o \ , \ \psi \in S^1, \qquad 5.7\text{-}4$$
$$\dot{y} = \epsilon X^{1o}(y) + \epsilon^2 X^{2o} + \epsilon^3 R_2(\psi, y; \epsilon), \quad y(0) = y_o \ , \ y \in D \subset \mathbf{R}^n.$$

Here $(\psi_o, y_o)$ is the solution of the equation

$$(\phi_o, x_o) = (\psi_o + \epsilon v^{(1)}(\psi_o, y_o), y_o + \epsilon u^{(1)}(\psi_o, y_o) + \epsilon^2 u^{(2)}(\psi_o, y_o)).$$

On $S^1 \times D \times (0, \epsilon_o]$ we have the following (nonuniform) estimates for $R_1$ and $R_2$:

$$R_1 = O(\frac{1}{\Omega^3(y)}),$$
$$R_2 = O(\frac{1}{\Omega^4(y)})$$

and $v^{(1)}$, $u^{(1)}$ and $u^{(2)}$ can be estimated by

$$v^{(1)} = O(\frac{1}{\Omega^2(y)}),$$
$$u^{(1)} = O(\frac{1}{\Omega(y)}),$$
$$u^{(2)} = O(\frac{1}{\Omega^2(y)}).$$

$X^{1o}$ and $X^{2o}$ are defined as follows:

$$X^{1o}(y) = \int_{S^1} X(\phi, y) d\phi,$$

$$\Omega(y) u^{(1)}(\phi, y) = \int^{\phi} (X(\psi, y) - X^{1o}(y)) d\psi,$$

$$X^{2o}(y) = \int_{S^1} (\nabla X(\phi,y) - X(\phi,y)\frac{\nabla\Omega(y)}{\Omega(y)})u^{(1)}(\phi,y)d\phi.$$

It follows that

$$X^{1o}(y) = O(1), \qquad X^{2o}(y) = O(\frac{1}{\Omega^2(y)}).$$

Of course, one has to choose $(\phi_o, x_o)$ well outside the inner domain, since otherwise it may not be possible to solve the equation for the initial conditions $(\psi_o, y_o)$. In the same sense the estimates are non-uniform. For the proof to work, one has to require that

$$\frac{\epsilon}{\Omega^2(y)} = o(1) \text{ as } \epsilon \downarrow 0.$$

That is to say, $y$ should be outside a $\epsilon^{\frac{1}{2}}$- neighborhood of the resonance manifold.

**Proof**

First we differentiate the relation 5.7-3 along the vectorfield:

$$\begin{bmatrix}\dot{\phi}\\ \dot{x}\end{bmatrix} = \begin{bmatrix} 1+\epsilon\frac{\partial v^{(1)}}{\partial\psi} & \epsilon\nabla v^{(1)} \\ \epsilon\frac{\partial u^{(1)}}{\partial\psi}+\epsilon^2\frac{\partial u^{(2)}}{\partial\psi} & 1+\epsilon\nabla u^{(1)}+\epsilon^2\nabla u^{(2)} \end{bmatrix} \begin{bmatrix}\dot{\psi}\\ \dot{y}\end{bmatrix}$$

$$= \begin{bmatrix} 1+\epsilon\frac{\partial v^{(1)}}{\partial\psi} & \epsilon\nabla v^{(1)} \\ \epsilon\frac{\partial u^{(1)}}{\partial\psi}+\epsilon^2\frac{\partial u^{(2)}}{\partial\psi} & 1+\epsilon\nabla u^{(1)}+\epsilon^2\nabla u^{(2)} \end{bmatrix} \begin{bmatrix}\Omega(y)+\epsilon^2 R_1\\ \epsilon X^{1o}+\epsilon^2 X^{2o}+\epsilon^3 R_2\end{bmatrix}$$

$$= \begin{bmatrix}\Omega(y)\\ 0\end{bmatrix} + \epsilon\begin{bmatrix}\frac{\partial v^{(1)}}{\partial\psi}\Omega(y)\\ \frac{\partial u^{(1)}}{\partial\psi}\Omega(y)+X^{1o}\end{bmatrix} + \epsilon^2\begin{bmatrix}R_1+\nabla v X^{1o}\\ \frac{\partial u^{(2)}}{\partial\psi}\Omega+X^{2o}+\nabla u^{(1)}X^{1o}\end{bmatrix}$$

$$+\epsilon^3\begin{bmatrix}\frac{\partial v^{(1)}}{\partial\psi}R_1+\nabla v^{(1)}X^{2o}\\ \frac{\partial u^{(1)}}{\partial\psi}R_1+\nabla u^{(2)}X^{1o}+\nabla u^{(1)}X^{2o}+R_2\end{bmatrix} + \epsilon^4\begin{bmatrix}\nabla v^{(1)}R_2\\ \frac{\partial u^{(2)}}{\partial\psi}R_1+\nabla u^{(2)}X^{2o}+\nabla u^{(1)}R_2\end{bmatrix}$$

$$+\epsilon^5\begin{bmatrix}0\\ \nabla u^{(2)}R_2\end{bmatrix}.$$

Then we use 5.7-3 to replace $(\phi, x)$ by $(\psi, y)$ in the original differential equation:

$$\begin{bmatrix}\dot{\phi}\\ \dot{x}\end{bmatrix}=\begin{bmatrix}\Omega(x)\\ \epsilon X(\phi,x)\end{bmatrix}=\begin{bmatrix}\Omega(y+\epsilon u^{(1)}+\epsilon^2 u^{(2)})\\ \epsilon X(\phi+\epsilon v^{(1)},y+\epsilon u^{(1)}+\epsilon^2 u^{(2)})\end{bmatrix}=$$

$$=\begin{bmatrix}\Omega(y)+\epsilon\nabla\Omega u^{(1)}+O(\epsilon^2(u^{(1)})^2+\epsilon^2 u^{(2)})\\ \epsilon X(\phi,y)+\epsilon^2\frac{\partial X}{\partial\phi}v^{(1)}+\epsilon^2\nabla X u^{(1)}+O(\epsilon^3((u^{(1)})^2+u^{(2)}))\end{bmatrix}=$$

$$=\begin{bmatrix}\Omega(y)\\ 0\end{bmatrix}+\epsilon\begin{bmatrix}\nabla\Omega u^{(1)}\\ X(\phi,y)\end{bmatrix}+\epsilon^2\begin{bmatrix}O((u^{(1)})^2+u^{(2)})\\ \frac{\partial X}{\partial\phi}v^{(1)}+\nabla X u^{(1)}\end{bmatrix}+\epsilon^3\begin{bmatrix}0\\ O((u^{(1)})^2)+u^{(2)})\end{bmatrix}.$$

Equating powers of $\epsilon$, we obtain the following relations:

$$\begin{bmatrix}\frac{\partial v}{\partial\psi}^1\\ \frac{\partial u}{\partial\psi}^1\end{bmatrix}\Omega(y)=\begin{bmatrix}\nabla\Omega u^{(1)}\\ X-X^{1o}\end{bmatrix},$$

$$\frac{\partial u}{\partial\psi}^2\Omega=\frac{\partial X}{\partial\psi}v^{(1)}+\nabla X u^{(1)}-\nabla u^{(1)}X^{1o}-X^{2o}.$$

The second component of the first equation is by now standard: Let $u^{(1)}$ be defined by

$$u^{(1)}(\psi,y)=\frac{1}{\Omega(y)}\int^{\psi}\{X(\phi,y)-X^{1o}(y)\}d\phi$$

and

$$\int_{S^1}u^{(1)}d\psi=0,$$

where, as usual

$$X^{1o}(y)=\int_{S^1}X(\phi,y)d\phi.$$

Then we can also solve the first component (if the average of $u^{(1)}$ had not been zero, the averaged vectorfield would have been different):

$$v^{(1)}(\psi,y)=\frac{1}{\Omega(y)}\int^{\psi}\nabla\Omega(y)u^{(1)}(\phi,y)d\phi$$

and again

$$\int_{S^1}v^{(1)}d\phi=0.$$

Thus we have that $u^{(1)}=O(\frac{1}{\Omega})$ and $v^{(1)}=O(\frac{1}{\Omega^2})$. We are now ready to solve the second equation:

$$u^{(2)}(\psi)=\frac{1}{\Omega}\int^{\psi}[\frac{\partial X}{\partial\phi}(\phi)v^{(1)}(\phi)+\nabla X(\phi)u^{(1)}(\phi)-\nabla u^{(1)}(\phi)X^{1o}-X^{2o}]d\phi.$$

This is a bounded expression if we take

$$X^{2o}(y)=\int_{S^1}[\frac{\partial X}{\partial\phi}v^{(1)}(\phi,y)+\nabla X(\phi,y)u^{(1)}(\phi,y)-\nabla u^{(1)}(\phi,y)X^{1o}(y)]d\phi$$

$$=\int_{S^1}[\frac{\partial X}{\partial\phi}v^{(1)}(\phi,y)+\nabla X(\phi,y)u^{(1)}(\phi,y)]d\phi$$

$$=\int_{S^1}[\nabla X(\phi,y)u^{(1)}(\phi,y)-X(\phi,y)\frac{\partial v^{(1)}}{\partial\phi}(\phi,y)]d\phi$$

$$=\int_{S^1}[\nabla X(\phi,y)u^{(1)}(\phi,y)-\frac{\nabla\Omega(y)}{\Omega(y)}X(\phi,y)u^{(1)}(\phi,y)]d\phi.$$

From this last expression it follows that we do not have to compute either $v^{(1)}$ or $u^{(2)}$, in order to compute $X^{1o}$ and $X^{2o}$. It also follows that $X^{2o}=O(\frac{1}{\Omega^2})$. We can solve $R_1$ and $R_2$ from the equations, and we find that $R_1=O(\frac{1}{\Omega^4})$ and $R_2=O(\frac{1}{\Omega^3})$. In each expansion we have implicitly assumed that $\frac{\epsilon}{\Omega^2}=o(1)$ as $\epsilon\downarrow 0$.

### 5.7.5. Exercise

Formulate and prove the analogue of Lemma 5.3.2, for the situation described in Lemma 5.7.4. □

In Lemma 5.3.2, one restriction is that $\Omega$ be bounded away from zero in a uniform way. We shall have to drop this restriction, and assume for instance that the distance of the boundary of the outer domain to the resonance manifold is at least of order $\delta(\epsilon)$, where $\delta(\epsilon)$ is somewhere between $\epsilon^{1/2}$ and 1.

Then there is the time-scale. If there is no passage through resonance, i.e. if in Lemma 5.7.1 the average of $X_+$ vanishes, there is no immediate need to prove anything on a longer time-scale than $\frac{1}{\epsilon^{1/2}}$, since that is the natural time-scale in the inner region (this, of course, is a matter of taste; one might actually need longer time-scale estimates in the outer region: the reader may want to try to formulate a lemma on this problem, cf.(San78a). On the other hand, if there is a passage through resonance, then we can use this as follows:

In our estimate, we meet expressions of the form

$$\int\frac{\epsilon}{\Omega^k(z(t))}dt,$$

where $z$ is the solution of the outer equation

$$\dot{z}=\epsilon X^{1o}(z)+\epsilon^2X^{2o}(z).$$

Let us take a simple example:

$$\dot{\phi}=z,$$

$$\dot{z} = \epsilon\alpha.$$

Then $z(t) = z_o + \epsilon\alpha t$ and the integral is

$$\int_{t_1}^{t_2} \frac{\epsilon dt}{(z_o + \epsilon\alpha t)^k} = \frac{1}{\alpha}\int_{z(t_1)}^{z(t_2)} \frac{d\xi}{\xi^k} = \frac{1}{\alpha(k-1)}\left(\frac{1}{z^{k-1}(t_1)} - \frac{1}{z^{k-1}(t_2)}\right) =$$

$$= \frac{1}{\alpha(k-1)}\left(\frac{1}{\Omega^{k-1}(t_1)} - \frac{1}{\Omega^{k-1}(t_2)}\right), \; k \geqslant 2.$$

This leads to the following

### 5.7.6. Assumption

In the sequel we shall assume that for $k \geqslant 2$

$$\int_{t_1}^{t_2} \frac{\epsilon dt}{\Omega^k(z(t))} = O\left(\frac{1}{\Omega^{k-1}(z(t_1))}\right) + O\left(\frac{1}{\Omega^{k-1}(z(t_2))}\right)$$

as long as $z(t)$ stays in the outer domain on $[t_1, t_2]$.

This is an assumption that can be checked, at least in principle, since it only involves the averaged equation, which we are supposed to be able to solve (Actually, solving may not even be necessary; all we need is some nice estimates on $\Omega(z(t))$).

One might wish to prove this assumption from basic facts about the vectorfield, but this turns out to be difficult, since this estimate incorporates rather subtly both the velocity of the $z$-component, and the dependence of $\Omega$ on $z$.

### 5.7.7. Lemma

Consider equation 5.7-4, introduced in Lemma 5.7.4,

$$\dot{\psi} = \Omega(y) + \epsilon^2 R_1(\psi, y; \epsilon), \qquad \psi(0) = \psi_o \,,\; \psi \in S^1,$$

$$\dot{y} = \epsilon X^{1o}(y) + \epsilon^2 X^{2o} + \epsilon^3 R_2(\psi, y; \epsilon), \;\; y(0) = y_o \,,\; y \in D \subset \mathbf{R}^n,$$

where we have the following estimates

$$R_1 = O\left(\frac{1}{\Omega^3(y)}\right), \; R_2 = O\left(\frac{1}{\Omega^4(y)}\right), \; \textit{and } X^{2o} = O\left(\frac{1}{\Omega^2(y)}\right).$$

Let $(\psi, y)$ be the solution of this equation. Let $(\zeta, z)$ be the solution of the truncated system

$$\dot{\zeta} = \Omega(z), \qquad \zeta(0) = \zeta_o \,,\; \zeta \in S^1,$$

$$\dot{z} = \epsilon X^{1o}(z) + \epsilon^2 X^{2o}(z) \,,\; z(0) = z_o \,,\; z \in D^o \subset D \subset \mathbf{R}^n$$

and assume that for $z$ in the outer region and $k \geqslant 2$

$$\int_{t_1}^{t_2} \frac{\epsilon dt}{\Omega^k(z(t))} = O\left(\frac{1}{\Omega^{k-1}(z(t_1))}\right) + O\left(\frac{1}{\Omega^{k-1}(z(t_2))}\right).$$

Then $(\zeta,z)$ is an approximation of $(\psi,y)$ in the following sense: Let $\delta$ be such that $\frac{\epsilon}{\delta^2(\epsilon)}=o(1)$ and $|\Omega(z(t))|\geqslant C\delta(\epsilon)$ for all $t\in[0,\frac{L}{\epsilon})$; then on $0\leqslant\epsilon t\leqslant L$

$$\|y(t)-z(t)\|=O(\|y_o-z_o\|)+O(\frac{\epsilon^2}{\delta_3(\epsilon)}),$$

$$|\psi(t)-\zeta(t)|=O(|\psi_o-\zeta_o|)+O(\frac{\|y_o-z_o\|}{\epsilon})+O(\frac{\epsilon}{\delta^2(\epsilon)}).$$

**Proof**

In the following estimates, we shall not go into all technical details; the reader is invited to plug any holes he might notice.

Using the differential equations, we obtain the following estimate for the difference between $y$ and $z$:

$$\|y(t)-z(t)\|\leqslant\|y_o-z_o\|+\epsilon\int_0^t\|X^{1o}(y(\tau))-X^{1o}(z(\tau))\|d\tau$$

$$+\epsilon^2\int_0^t\|X^{2o}(y(\tau))-X^{2o}(z(\tau))\|d\tau+\epsilon^3\int_0^t\|R_2(\psi(\tau),y(\tau);\epsilon)\|d\tau$$

$$\leqslant\|y_o-z_o\|+\epsilon\int_0^t C\{1+\frac{\epsilon}{\Omega^3(z(\tau))}+\frac{\epsilon^2}{\Omega^5(z(\tau))}\}\|y(\tau)-z(\tau)\|d\tau+\epsilon^3\int_0^t\frac{C}{\Omega^4(z(\tau))}d\tau$$

(For odd powers of $\Omega$, we take of course the power of the absolute value). Using the *Gronwall* Lemma, this implies

$$\|y(t)-z(t)\|\leqslant\|y_o-z_o\|e^{\epsilon\int_0^t C\{1+\frac{\epsilon}{\Omega^3(z(\tau))}+\frac{\epsilon^2}{\Omega^5(z(\tau))}\}d\tau}$$

$$+\int_0^t\frac{C\epsilon^3}{\Omega^4(z(\tau))}e^{\epsilon\int_0^t C\{1+\frac{\epsilon}{\Omega^3(z(\sigma))}+\frac{\epsilon^2}{\Omega^5(z(\sigma))}\}d\sigma}d\tau$$

$$\leqslant(\|y_o-z_o\|+C(\frac{\epsilon^2}{\Omega^3(z(t))}+\frac{\epsilon^2}{\Omega^3(z(0))}))$$

$$e^{C(\epsilon t+\frac{\epsilon}{\Omega^2(z(t))}+\frac{\epsilon}{\Omega^2(z(0))}+\frac{\epsilon^2}{\Omega^4(z(t))}+\frac{\epsilon^2}{\Omega^4(z(0))})}$$

$$=O(\|y_o-z_o\|)+O(\frac{\epsilon^2}{\delta^3})\ on\ 0\leqslant\epsilon t\leqslant L.$$

For the angular variables, the estimate is now very easy:

$$\|\psi(t)-\zeta(t)\|\leqslant\|\psi_o-\zeta_o\|+C\int_0^t\|z(\tau)-y(\tau)\|d\tau+C\epsilon^2\int_0^t\frac{1}{\Omega^3(z(\tau))}d\tau$$

$$=O(\|\psi_o-\zeta_o\|)+O(\frac{\|y_o-z_o\|}{\epsilon})+O(\frac{\epsilon}{\delta^2}) \text{ on } 0\leq\epsilon t\leq L$$

if $d(z_o,N)=O_s(1)$; otherwise one has to include a term $O(\frac{\epsilon}{\Omega^3(z_o)})$; this last term presents difficulties if one wishes to obtain estimates for the full passage through resonance, at least for the angular variables.

The estimate implies that one can average as long as $\frac{\epsilon}{\delta^2}=o(1)$. On the other hand, the estimates for the inner region are only valid in a $\epsilon^{\frac{1}{2}}$-neighborhood of the resonance manifold. So there is a gap. As we have already pointed out however, it is possible to bridge this gap by using a time-scale extension argument for one-dimensional monotone vectorfields. The full proof of this statement is rather complicated and has been given in (San79a).

Here we would like to give only the essence of the argument in the form of a lemma. This lemma is due to *Eckhaus* and has not been published before.

**5.7.8. Lemma (*Eckhaus*)**

Consider the vectorfield

$$\dot{x}=f(x)+\epsilon R(t,x)\ ,\ x(0)=x_o\ ,\ x\in D\subset\mathbf{R},$$

with $0<m\leq f(x)\leq M<\infty$, $\|R\|\leq C$ for $x\in D$. Then $y$, the solution of

$$\dot{y}=f(y)\ ,\ y(0)=x_o$$

is an $O(\epsilon t)$-approximation of $x$ (as compared to the usual $O(\epsilon te^{\epsilon t})$ *Gronwall*-estimate).

**Proof**

Let $y$ be the solution of

$$\dot{y}=f(y)\ ,\ y(0)=x_o$$

and let $t^*$ be the solution of

$$\dot{t}^*=1+\epsilon\frac{R(t,y(t^*))}{f(y(t^*))}\ ,\ t^*(0)=0.$$

This equation is well defined and we have

$$|y(t^*(t))-y(t)|\leq\int_t^{t^*(t)}|\dot{y}(\tau)|\,d\tau\leq M\,|t^*(t)-t|$$

while on the other hand

$$|t^*(t)-t|\leq\int_0^t\epsilon\,|\frac{R(\tau,y(t^*(\tau)))}{f(y(t^*(\tau)))}|\,d\tau\leq\frac{C}{m}\epsilon t.$$

Therefore $|y(t^*(t))-y(t)|\leq C\frac{M}{m}\epsilon t$. Let

$$\tilde{x}(t)=y(t^*(t)).$$

Then $\tilde{x}(0)=y(t^*(0))=y(0)=x_o$ and

$$\dot{\tilde{x}}(t)=\dot{t}^*\dot{y}=(1+\epsilon\frac{R(t,y(t^*(t)))}{f(y(t^*(t))})f(y(t^*(t)))$$

$$=f(y(t^*(t)))+\epsilon R(t,y(t^*(t)))$$

$$=f(\tilde{x}(t))+\epsilon R(t,\tilde{x}(t)).$$

By uniqueness, $\tilde{x}=x$, the solution of the original equation, and therefore $x(t)-y(t)=O(\epsilon t)$. □

Although we have a higher dimensional problem, the inner equation is only two-dimensional and integrable. This makes it possible to apply a variant of this lemma in our situation. The nice thing about this lemma is that it gives explicit order estimates and we do not have to rely on abstract extension principles giving only $o(1)$-estimates.

**Concluding remarks**

Although we have neither given the full theorem on passage through resonance, nor an adequate discussion of the technical difficulties, we have given here the main ideas and concepts that are needed to do this.

Note that from the point of view of asymptotics passage through resonance and locking in resonance is still an open problem. The main difficulty arises as follows: In the case of locking the solution in the inner domain enters a ball in which all solutions attract toward a critical point or periodic solution. During the start of this process the solution has to pass fairly close to the saddle point (since the inner equation is conservative to first order, all attraction has to be small, and so is the splitting up of the stable and unstable manifold). While passing the saddle point, errors grow like $e^{\epsilon^{1/2}t}$ ($\frac{1}{\epsilon^{1/2}}$ being the natural time scale in the inner domain); the attraction in the ball on the other hand takes place on the time-scale $\frac{1}{\epsilon}$ so that we get into trouble with the asymptotic estimates in the case of attraction (see the preceding chapter).

Finally it should be mentioned that the method of multiple time-scales provides us with the same formal results but equally fails to describe the full asymptotic problem rigorously.

## 5.8. Two examples

### 5.8.1. The forced mathematical pendulum

We shall sketch the treatment of the perturbed pendulum equation

$$\ddot{\phi}+\sin(\phi)=\epsilon F(t,\phi,\dot{\phi})$$

while specifying the results in the case $F=\sin(t)$. The treatment is rather technical, involving elliptic integrals and the details can be found in appendix 8.3. The notation of elliptic integrals and a number of basic results are taken from (Byr71a). This calculation has been inspired by (Gre84a) and (Cap73a) and some conversations with *S. − N.Chow*.

Putting $\epsilon=0$ the equation has the energy integral

$$\tfrac{1}{2}\dot{\phi}^2-\cos(\phi)=c.$$

It is convenient to introduce $\phi=2\theta$ and $k^2=\frac{2}{1+c}$; then

$$\dot{\theta}=\pm\frac{1}{k}(1-k^2\sin^2(\theta))^{\frac{1}{2}}.$$

Instead of $t$ we introduce the time-like variable $u$ by

$$t=k\int^{\theta}\frac{d\tau}{(1-k^2\sin^2\tau)^{\frac{1}{2}}}=ku.$$

This implies $\sin(\theta)=sn(ku,k)$ and

$$\dot{\theta}=\pm\frac{1}{k}(1-k^2sn^2(ku,k))^{\frac{1}{2}}=\pm\frac{1}{k}dn(ku,k).$$

In the spirit of the method of variation of constants we introduce for the perturbed problem the transformation $(\theta,\dot{\theta})\mapsto(k,u)$.

$$\theta=am(ku,k),$$

$$\dot{\theta}=\frac{1}{k}dn(ku,k).$$

After some manipulation of elliptic functions we find

$$\dot{k}=-\frac{\epsilon}{2}k^2dn(ku,k)F,$$

$$\dot{u}=\frac{1}{k^2}+\frac{\epsilon}{2}\frac{F}{1-k^2}(-E(ku)dn(ku,k)+k^2sn(ku,k)cn(ku,k)).$$

One can demonstrate that

$$\dot{u}=\frac{1}{k^2}+O(\epsilon)\ \textit{uniform in } k.$$

If $F$ depends explicitly and periodically on time, the system of equations for $k$ and $u$ constitutes an example which can be handled by averaging over two angles $t$ and $u$, except that one has to be careful with the infinite series. We take

$$F=\sin(t).$$

*Fourier* expansion of the right hand side of the equation for $k$ can be written down using the elliptic integral $K(k)$:

$$\dot{k} = \frac{\epsilon k^2 \pi}{4K(k)} \sum_{m=-\infty}^{\infty} \frac{1}{\cosh \frac{m\pi K'(k)}{K(k)}} \sin\left(\frac{m\pi k}{K(k)}u - t\right).$$

It seems evident that we should introduce the angles

$$\psi_m = \frac{m\pi k}{K(k)}u - t,$$

with corresponding equation

$$\dot{\psi}_m = \frac{m\pi}{kK(k)} - 1 + O(\epsilon).$$

A resonance arises if for some $m = m_r$, and certain $k$

$$\frac{m_r \pi}{kK(k)} = 1.$$

We call the resonant angle $\psi_r$. The analysis until here runs along the lines pointed out in § 5.6 ; there are only technical complications owing to the use of elliptic functions. We have to use an averaging transformation which is a direct extension of the case with one angle. It takes the form

$$x = y + \epsilon \sum_{m \neq m_r} u_m(\psi_m, y)$$

For details see appendix 8.3.

As a result of the calculations we have the following vectorfield in the $m_r$-th resonance domain

$$\dot{k} = \epsilon \frac{k^2}{4K(k)} [\cosh(\frac{m_r \pi K'}{K})]^{-1} \sin(\psi_r),$$

$$\dot{\psi}_r = \frac{m_r \pi}{kK(k)} - 1 + O(\epsilon).$$

This means that for $k$ corresponding with the resonance given by $m = m_r$ there are two periodic solutions, given by $\psi_r = 0, \pi$, one elliptic, one hyperbolic. Note that we can formally take the limit for $m_r \to \infty$ while staying in resonance:

$$\lim_{\substack{m_r \to \infty \\ m_r \pi = kK(k)}} \dot{k} = \frac{\epsilon \pi}{4K(k) \cosh(\frac{\pi}{2})} \sin(\psi_r)$$

One should compare this answer with the *Melnikov* function (see appendix 8.4) for this particular problem. (See also (San82a)).

**5.8.2. An oscillator attached to a fly-wheel**

We return to example 1.1.6, which describes a fly-wheel mounted on an elastic support; the description was taken from (Gol71a), chapter 8.3. The same model has been described in (Eva76a), chapter 3.3 where one studies a motor with a slightly eccentric rotor, which interacts with its elastic support. First we obtain the equations of motion for $x$, the oscillator, and $\phi$, the rotational motion, separately. We abbreviate $M_1(\dot{\phi})=M(\dot{\phi})-M_w(\dot{\phi})$, obtaining

$$\ddot{x}+\omega_o^2 x=\frac{\epsilon}{m}\frac{-f(x)-\beta\dot{x}+q_1\dot{\phi}^2\cos(\phi)}{1-\epsilon^2\frac{q_1q_2}{m}\sin^2\phi}+\epsilon^2R_1,$$

$$\ddot{\phi}=\frac{\frac{\epsilon}{J_o}M_1(\dot{\phi})+\epsilon q_2 g\sin(\phi)-\epsilon q_2\omega_o^2 x\sin(\phi)}{1-\epsilon^2\frac{q_1q_2}{m}\sin^2(\phi)}+\epsilon^2R_2,$$

with

$$R_1=\frac{-\frac{q_2\omega_o^2}{m}x\sin^2(\phi)+\frac{q_1}{m}\sin(\phi)(\frac{1}{J_o}M_1(\dot{\phi})+q_2 g\sin(\phi)}{1-\epsilon^2\frac{q_1q_2}{m}\sin^2(\phi)},$$

$$R_2=\frac{\frac{q_2}{m}\sin(\phi)(-f(x)-\beta\dot{x}+q_1\dot{\phi}^2\cos(\phi))}{1-\epsilon^2\frac{q_1q_2}{m}\sin^2(\phi)}.$$

Using the transformation from § 1.1, $x=r\sin(\phi_2)$, $\dot{x}=\omega_o r\cos(\phi_2)$, $\phi=\phi_1$, $\dot{\phi}=\Omega$ we find to first order

$$\dot{r}=\epsilon\frac{\cos(\phi_2)}{m\omega_o}(-f(r\sin(\phi_2)-\beta\omega_o r\cos(\phi_2)+q_1\Omega^2\cos(\phi_1))+O(\epsilon^2),$$

$$\dot{\Omega}=\frac{\epsilon}{J_o}M_1(\Omega)+\epsilon q_2 g\sin(\phi_1)-\epsilon q_2\omega_o^2 r\sin(\phi_1)\sin(\phi_2)+O(\epsilon^2),$$

$$\dot{\phi}_1=\Omega, \qquad 5.8.2\text{-}1$$

$$\dot{\phi}_2=\omega_o+O(\epsilon).$$

The right hand sides of equation 5.8.2-1 can be written as separate functions of $\phi_2$, $\phi_1-\phi_2$ and $\phi_1+\phi_2$. Assuming $\Omega$ to be positive we have only one resonance manifold given by

$$\Omega=\omega_o.$$

Outside the resonance we average over the angles to obtain

$$\dot{r}=-\epsilon\frac{\beta}{2m}r+O(\epsilon^2), \qquad 5.8.2\text{-}2$$

$$\dot{\Omega} = \frac{\epsilon}{J_o} M_1(\Omega) + O(\epsilon^2).$$

Depending on the choice of the motor characteristic $M_1(\Omega)$, the initial value of $\Omega$ and the eigenfrequency $\omega_o$ the system will move into resonance or stay outside the resonance domain. In the resonance domain near $\Omega = \omega_o$ averaging over the angles $\phi_2$ and $\phi_1 + \phi_2$ produces

$$\dot{r} = -\epsilon \frac{\beta}{2m} r + \epsilon \frac{q_1}{2m\omega_o} \Omega^2 \cos(\phi_1 - \phi_2) + O(\epsilon^2),$$

$$\dot{\Omega} = \frac{\epsilon}{J_o} M_1(\Omega) - \epsilon \frac{q_2 \omega_o^2}{2} r \cos(\phi_1 - \phi_2) + O(\epsilon^2),$$

$$\dot{\phi}_1 - \dot{\phi}_2 = \Omega - \omega_o + O(\epsilon).$$

Putting $\chi = \phi_1 - \phi_2$ we can derive the inner equation

$$\ddot{\chi} = \frac{\epsilon}{J_o} M_1(\Omega) - \epsilon \frac{q_2 \omega_o^2}{2} r \cos(\chi),$$

$$\dot{r} = 0.$$

To study the possibility of locking into resonance we have to analyze the equilibrium solutions of the equation for $r$, $\Omega$ and $\chi$ in the resonance domain. If we find stability we should realize that we cannot expect the equilibrium solutions to be globally attracting. Some solutions will be attracted into the resonance domain and stay there, others will pass through the resonance.

The equilibrium solutions are given by

$$\frac{\beta}{2m} r_* = \frac{q_1}{2m\omega_o} \Omega_*^2 \cos(\chi_*),$$

$$\frac{1}{J_o} M_1(\Omega_*) = \frac{q_2 \omega_o^2}{2} r_* \cos(\chi_*),$$

$$\Omega_* = \omega_o.$$

The analysis produces three small eigenvalues containing terms of $O(\epsilon^{1/2})$, $O(\epsilon)$ and higher order. A second order approximation of the equations of motion and the eigenvalues may be advisable but we do not perform this computation here (Note that in (Gol71a), page 319, a force $P(x) = -cx - \gamma x^3$ is used which introduces a mixture of first- and second-order terms; from the point of asymptotics this is not quite satisfactory).

We conclude our study of the first-order calculation by choosing the constants and $M_1(\Omega)$ explicitly; this enables us to compare against numerical results. Choose $m = \omega_o = \beta = q_1 = q_2 = J_o = g = 1$; a linear representation is suitable for the motor characteristic:

$$M_1(\Omega) = \tfrac{1}{4}(2 - \Omega).$$

The equilibrium solutions are then given by

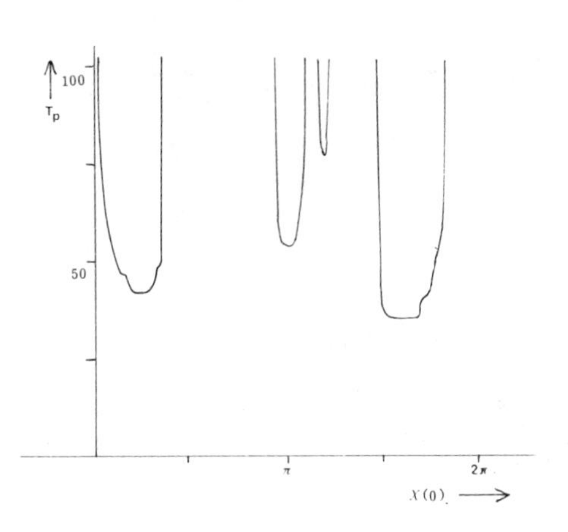


**Figure 5.8.2-1** *Time of passage through resonance $T_p$ as a function of the initial condition $\chi(0)$; $r(0)=1$, $\Omega(0)=0.5$, $\Omega(T_p)=1.5$.*

$$r_\star = \cos(\chi_\star),$$

$$r_\star \cos(\chi_\star) = \tfrac{1}{2},$$

$$\Omega_\star = 1$$

so that $r_\star = \frac{1}{2^{1/2}}$ and $\chi_\star = \pm\frac{\pi}{4}$. The calculation thus far suggests that locally, in the resonance manifold $r = \frac{1}{2^{1/2}}$, $\Omega = 1$, stable attracting solutions may exist corresponding with the phenomenon of locking in resonance. Starting with initial conditions $r(0) > \frac{1}{2^{1/2}}$, $\Omega(0) < 1$ equations 5.8.2-2 tell us that we move into resonance; some of the solutions, depending on the initial value of $\chi$, will be caught in the resonance domain; other solutions will pass through resonance, and, again according to equation 5.8.2-2, will move to a region where $r$ is near zero and $\Omega$ is near 2. We illustrate this in figure 5.8.2-1 where the time of passage through resonance is plotted against the initial values.

# 6. Normal Forms

## 6.1. Introduction

In this chapter we shall consider the idea of *normal form* in the context of averaging. Loosely speaking, a mathematical object, be it a matrix, a function or a vector field, to name but a few, is said to be in normal form if it is 'reasonably simple' and is obtained by coordinate transformations. There are two reasons for this last requirement:

1. Properties of mathematical objects should not depend on the choice of coordinates, so if one transforms one object onto another by a coordinate transformation, the two are equivalent.

2. Coordinate transformations are a good computational tool.

The first requirement, that it be simple, is not so clear-cut. In practice this might be useful as a guiding principle, but the exact definition is as much a matter of what one wants (maximal simplicity) as of what one can get.

A good illustration is found in linear algebra. Here the *Jordan* normal form is found by applying linear transformations to the space on which a linear operator works in order to put this operator in (almost) diagonal form. While the diagonal terms are determined by the original operator, the off-diagonal terms depend on the chosen transformation. The same argument applies to linear differential equations, and it is only a small step to drop the linearity. In the next part we shall try to define what a normal form should look like, without using transformations explicitly, but, of course, relying on our experience with averaging. Note that the theory of normal forms goes back to at least as early as *Euler*; *Delaunay* , *Poincaré*, *Dulac* and *Birkhoff* contributed towards a more definitive form of the theory. A large amount of this work is concerned with near-identity transformations of systems like

$$\dot{x} = Ax + F(x)$$

where $A$ is a $n \times n$-matrix, $F(x)$ stands for higher order (nonlinear) terms in $x$; See e.g. *Arnol'd* (Chapter 5 of (Arn78a)).

**Motivation**

Consider as an example the differential equation (periodic in $t$)

$$\dot{x} = \epsilon f(t,x).$$

We write this equation as

$$\dot{x} = \epsilon f(\phi,x),$$

$$\dot{\phi} = 1.$$

We can associate with this system the following operator:

$$S + \epsilon P = \frac{\partial}{\partial\phi} + \epsilon f \nabla.$$

The connection between the two is as follows: if we have a function of $\phi$ and $x$ then we get the same result if we apply the operator or differentiate along the vectorfield (take the total time derivative). In general f depends on $\phi$ but we say that f is in normal form if it is time-independent (equal to its average). Since

$$[S,P]\chi = \frac{\partial}{\partial\phi} f \nabla\chi - f\nabla\frac{\partial}{\partial\phi}\chi = \frac{\partial f}{\partial\phi}\nabla\chi + f\frac{\partial}{\partial\phi}\nabla\chi - f\nabla\frac{\partial}{\partial\phi}\chi = \frac{\partial f}{\partial\phi}\nabla\chi$$

we find that

$$\frac{\partial}{\partial\phi} f = 0 \Leftrightarrow [S,P] = 0.$$

If we let $adS : P \mapsto [S,P]$ , we say that the differential equation associated with $S$ and $P$ is in normal form if

$$P \in ker(adS).$$

This formulation corresponds with the usual definition for systems in the standard form

$$\dot{x} = \epsilon f(t,x)$$

and, as we shall show, also for Hamiltonian systems of the type

$$H = H_0(p) + \epsilon H_1(q,p).$$

There is, however, a difficulty with the definition which we introduced. Consider the system

$$\dot{\phi} = \Omega(x),$$

$$\dot{x} = \epsilon X(\phi,x).$$

Let $S+\epsilon P=\Omega(x)\frac{\partial}{\partial\phi} + \epsilon X(\phi,x)\nabla$. Then

$$[S,P]\chi = \Omega(x)\frac{\partial}{\partial\phi}X(\phi,x)\nabla\chi - X\nabla(\Omega(x)\frac{\partial}{\partial\phi}\chi)$$

$$= \Omega(x)\frac{\partial X}{\partial\phi}\nabla\chi + \Omega(x)X(\phi,x)\frac{\partial}{\partial\phi}\nabla\chi$$

$$- X(\nabla\Omega)\frac{\partial}{\partial\phi}\chi - X\Omega(x)\nabla\frac{\partial}{\partial\phi}\chi$$

$$= \Omega(x)\frac{\partial X}{\partial\phi}\nabla\chi - X(\nabla\Omega)\frac{\partial}{\partial\phi}\chi.$$

To apply our definition this has to be zero but it is not enough for X to be $\phi$ -independent, as we might have expected from the results in chapter 5, so we have to change our definition. One would expect the following definition to be a reasonable generalization:

**6.1.1. Definition:**

We say that the vector field

$$S + \epsilon P$$

is in normal form iff $P$ is in the kernel of the repeatedly applied commutator.

This turns out to be a useful definition, as we will see in the next section.

## 6.2. Definitions and examples

Let $\mathfrak{X}$ be the space of vector fields on some manifold M, and let $[\,,\,]$: $\mathfrak{X} \times \mathfrak{X} \rightarrow \mathfrak{X}$ be the *commutator* of the two vector fields, defined as

$$[X, Y] = L_X Y \qquad X, Y \in \mathfrak{X}.$$

In local coordinates, this is

$$[X, Y] = dY(X) - dX(Y),$$

since if $X = \sum_i X_i \frac{\partial}{\partial x_i}$ and $Y = \sum_i Y_i \frac{\partial}{\partial x_i}$,

$$[X, Y]\psi = \sum_i X_i \frac{\partial}{\partial x_i} \sum_j Y_j \frac{\partial}{\partial x_j}\psi - \sum_i Y_i \frac{\partial}{\partial x_i} \sum_j X_j \frac{\partial}{\partial x_j}\psi$$

$$= \sum_{i,j} X_i \frac{\partial}{\partial x_i} Y_j \frac{\partial}{\partial x_j}\psi + \sum_{i,j} X_i Y_j \frac{\partial^2}{\partial x_i \partial x_j}\psi$$

$$- \sum_{i,j} Y_i \frac{\partial}{\partial x_i} X_j \frac{\partial}{\partial x_j}\psi - \sum_{i,j} Y_i X_j \frac{\partial^2}{\partial x_i \partial x_j}\psi$$

$$= \sum_{i,j} (X_i \frac{\partial}{\partial x_i} Y_j - Y_i \frac{\partial}{\partial x_i} X_j) \frac{\partial}{\partial x_j}\psi$$

or

$$[X,Y]_i = \sum_j (X_j \frac{\partial}{\partial x_j} Y_i - Y_j \frac{\partial}{\partial x_j} X_i) = (dY(X) - dX(Y))_i.$$

We define

$$adX: \mathfrak{X} \to \mathfrak{X}$$

by

$$adX(Y) = [X,Y].$$

**6.2.1. Definition (See Introduction for motivation)**

Consider the vector field

$$S + \epsilon P.$$

This vector field is said to be in *normal form* iff there exists an $l \in \mathbf{N}$ such that

$$P \in ker(ad^l S).$$

This is the definition of normal form as a simple object. Another, more operational, definition might be: the vectorfield is in *normal form* if $P$ is in the complement of the image of *ad S*. For each class of problems one has to show that the operational definition is equivalent to the 'simple' definition. There are situations where this is certainly not the case, e.g. when $S(y)$ is linear in $\mathbf{R}^n$ with $n$ eigenvalues zero. In this case the complement of the image consists of the kernel of the *ad* of another operator, which can easily be derived from the finite dimensional irreducible representations of $\mathfrak{sl}(2,\mathbf{R})$ (see (Cus82a) or (Mey84a) ).

**6.2.2. Example**

Consider

$$\dot{x} = \epsilon f(t,x).$$

We can associate to this the operator:

$$\frac{\partial}{\partial t} + \epsilon f(t,x) \frac{\partial}{\partial x}.$$

Putting $S = \frac{\partial}{\partial t}$ and $P = f\frac{\partial}{\partial x}$, we find

$$[S,P]\psi = (SP - PS)\psi = \frac{\partial}{\partial t} f \frac{\partial}{\partial x}\psi - f\frac{\partial}{\partial x}\frac{\partial}{\partial t}\psi = \frac{\partial f}{\partial t}\frac{\partial \psi}{\partial x}.$$

If $f$ is time-independent ($f$ equals its average), then

$$P \in ker(adS).$$

We shall now consider the question whether one can put a vector field into normal form by a coordinate transformation, if only approximately:

Let

$$I + \epsilon U: M \rightarrow M$$

be such a near-identity transformation, with $U \in \mathfrak{X}$. In local coordinates, we have

$$\dot{x} = S(x) + \epsilon P(x), \quad x = y + \epsilon u(y).$$

The pull-back (i.e. the vectorfield obtained after the coordinate transformation) of $S + \epsilon P$ along $I + \epsilon U$ is easily computed to be

$$(I + \epsilon dU(y))\dot{y} = S(y + \epsilon U(y)) + \epsilon P(y)$$
$$= S(y) + \epsilon dS(U(y)) + \epsilon P(y) + O(\epsilon^2)$$

or

$$\dot{y} = S(y) + \epsilon(dS(y)(U(y)) - dU(y)(S(y)) + P(y)) + O(\epsilon^2)$$

and we find

$$(S + \epsilon P)_\star = S + \epsilon(P - adS(U)) + O(\epsilon^2)$$

(Where the $\star$ denotes the pull-back). Suppose we can split P as follows:

$$P = P^o + \tilde{P},$$

with

$$P^o \in ker(ad^l S) \quad \tilde{P} \in \mathrm{Im}(adS)$$

Then the obtained normal form would be

$$\dot{x} = S + \epsilon P^o$$

In concrete problems, the thing to be checked is this splitting. Here one has some freedom since the choice of l still has to be made. In practice, one would start with $l = 1$ and if the construction would not work, one increases $l$ until the desired result is obtained. Let us see what this means for our simple example:
In the T-periodic case there is a natural projection on $ker(adS)$ called averaging! We split f as follows: $f = f^o + \tilde{f}$, where

$$f^o = \frac{1}{T}\int_0^T f(t,x)dt.$$

Then we have to solve $adS(U) = \tilde{f}$, i.e.

$$\frac{\partial u}{\partial t} = \tilde{f}$$

and we can do this by putting

$$u(t,x) = \int^t \tilde{f}(\tau,x)d\tau$$

and u is then uniformly bounded. If f is not periodic, the boundedness of u has to be checked separately.
One could generalize the idea of normal form by relaxing the notion of

'coordinate transformation'. Some kind of requirement will have to be made with respect to the boundedness of u, however, for the asymptotics to make sense, cf *Balbi* (Bal82a).

## 6.3. Normal forms of vector fields with slowly varying frequency

In this section we consider systems of the following type:

$$\dot{\phi} = \Omega(x) + \epsilon Y(\phi,x) \qquad ,\phi \in S^1,$$

$$\dot{x} = \epsilon X(\phi,x) \qquad ,x \in \mathbf{R}^n.$$

Using our definition of normal form, we shall derive a normal form for this system. Our main interest here is theoretical, i.e. to put the treatment of this system in the framework of normalization; there will be small variations on the method developed in chapter 5 due to the appearance of Y in the equation for the angular velocity.
Let

$$S = \begin{pmatrix} \Omega \\ 0 \end{pmatrix}, \qquad P = \begin{pmatrix} Y \\ X \end{pmatrix}$$

and

$$U^{(1)} = P - adS(U^{(0)}).$$

Starting with $l = 1$ in the definition of normal form, we put $U^{(1)} = 0$ ; this enables us to solve for $U^{(0)}$ after which it is clear when $P$ is in normal form.
We make the following technical assumptions:
Let

$$\|\Omega\|_o^- = \sup_{x \in D} \frac{1}{|\Omega(x)|},$$

$$\|\Omega\|_1^- = \|\Omega\|_o^- + \min_{i=1,\cdots n} \|\frac{\partial \Omega}{\partial x_i}\|_o^-$$

and assume

$$\|\Omega\|_1^- \leq C$$

We have

$$P - adS(U^{(0)}) = \begin{pmatrix} Y \\ X \end{pmatrix} - \begin{pmatrix} \frac{\partial U_1^{(0)}}{\partial \phi} & \nabla U_1^{(0)} \\ \frac{\partial U_2^{(0)}}{\partial \phi} & \nabla U_2^{(0)} \end{pmatrix} \begin{pmatrix} \Omega \\ 0 \end{pmatrix} + \begin{pmatrix} 0 & \nabla \Omega \\ 0 & 0 \end{pmatrix} \begin{pmatrix} U_1^{(0)} \\ U_2^{(0)} \end{pmatrix}$$

$$= \begin{pmatrix} Y \\ X \end{pmatrix} - \Omega \frac{\partial U}{\partial \phi} + \begin{pmatrix} \nabla \Omega U_2^{(0)} \\ 0 \end{pmatrix}.$$

The first condition to solve the equation $P - adS(U^{(0)}) = 0$ is that $X$ has a vanishing average (i.e. $X^o = 0$ ). Then we can take

$$U_2^{(0)}(\phi,x) = \frac{1}{\Omega(x)} \int^{\phi} X(\psi,x)d\psi + c(x).$$

This leaves $\overline{U}_2^{(0)}$, the average of $U_2^{(0)}$, free to choose. Since $\min_{i=1,\cdots n} \| \frac{\partial\Omega}{\partial x_i} \|_{\overline{0}} \leqslant C$ there is a $j$ such that

$$\sup_{x \in D} \frac{1}{\left| \frac{\partial\Omega}{\partial x_j} \right|} \leqslant C$$

Let

$$\overline{U}_2^{(0)_k}(x) = \begin{cases} \frac{-Y^o(x)}{\frac{\partial\Omega}{\partial x_j}} & k = j \\ 0 & k \neq j \end{cases}$$

Then $\nabla\Omega\overline{U}_2^{(0)} + Y^o(x) = 0$ and we can solve

$$\Omega \frac{\partial}{\partial\phi} U_1^{(0)} = Y - Y^o$$

and the normal form equation is

$$\dot{\phi} = \Omega(x),$$

$$\dot{x} = 0.$$

The requirement on $\nabla\Omega$ is not a mere technical point; if it can not be met, the normal form equation would have to look like

$$\dot{\phi} = \Omega(x) + \epsilon Y^o(x),$$

$$\dot{x} = 0$$

by choosing $\overline{U}_2^{(0)} = 0$ .
This was the simple case, with $X^o = 0$, and it is already rather complicated. Let us now turn to the case $X^o \neq 0$ . We cannot solve the equation for $U^{(0)}$ with $U^{(1)} = 0$, so we require instead $U^{(1)} \in ker(adS)$:

$$U^{(1)} = P - adS(U^{(0)}),$$

$$0 = adS(U^{(1)})$$

This is equivalent to

$$U_1^{(1)} = Y - \Omega \frac{\partial}{\partial\phi} U_1^{(0)} + \nabla\Omega U_2^{(0)}, \qquad 6.3\text{-}1$$

$$U_2^{(1)} = X - \Omega \frac{\partial}{\partial\phi} U_2^{(0)}, \qquad 6.3\text{-}2$$

$$0 = \Omega \frac{\partial}{\partial \phi} U_1^{(1)} - \nabla \Omega U_2^{(1)}, \qquad 6.3\text{-}3$$

$$0 = \Omega \frac{\partial}{\partial \phi} U_2^{(1)}. \qquad 6.3\text{-}4$$

Solving 6.3-4, we obtain $U_2^{(1)} = U_2^{(1)}(x)$. To solve 6.3-2, we have to require

$$U_2^{(1)}(x) = X^o(x) = \int_{S^1} X(\phi,x) d\phi.$$

There is, however, an obstruction in equation 6.3-3: we cannot solve it unless

$$\nabla \Omega \cdot X^o(x) = 0.$$

In general, this will not be the case, so we extend the equation for the normal form:

$$U^{(1)} = P - adS \cdot U^{(o)},$$

$$U^{(2)} = adS \cdot U^{(1)},$$

$$0 = adS \cdot U^{(2)}$$

(reformulating the requirement that $P - adS.U^{(o)} \in ker(ad^2 S)$). This leads to

$$U_1^{(1)} = Y - \Omega \frac{\partial}{\partial \phi} U_1^{(0)} + \nabla \Omega \cdot U_2^{(0)}, \qquad 6.3\text{-}5$$

$$U_2^{(1)} = X - \Omega \frac{\partial}{\partial \phi} U_2^{(0)}, \qquad 6.3\text{-}6$$

$$U_1^{(2)} = \Omega \frac{\partial}{\partial \phi} U_1^{(1)} - \nabla \Omega \cdot U_2^{(1)}, \qquad 6.3\text{-}7$$

$$U_2^{(2)} = \Omega \frac{\partial}{\partial \phi} U_2^{(1)}, \qquad 6.3\text{-}8$$

$$0 = \Omega \frac{\partial}{\partial \phi} U_1^{(2)} - \nabla \Omega U_2^{(2)}, \qquad 6.3\text{-}9$$

$$0 = \Omega \frac{\partial}{\partial \phi} U_2^{(2)}. \qquad 6.3\text{-}10$$

Equation 6.3-10 leads to $U_2^{(2)} = U_2^{(2)}(x)$, but we can only solve 6.3-8 if $U_2^{(2)} = 0$. Next we solve 6.3-9 to obtain $U_1^{(2)} = U_1^{(2)}(x)$ and from 6.3-8 we conclude that $U_2^{(1)} = U_2^{(1)}(x)$. Again, it follows from 6.3-6 that $U_2^{(1)}(x) = X^o(x)$ and $U_1^{(2)}$ can be obtained from 6.3-7 by taking $U_1^{(2)}(x) = -\nabla \Omega(x) X^o(x)$. The solution of 6.3-5 follows the same construction as in the case $X^o = 0$, and we have solved the normal form equation.

## 6.4. Normalization of Hamiltonian systems

In this section we give the classical theory of *Hamiltonian* normal forms using generating functions. This approach is rather awkward in its actual use, due to the fact that the generating function is defined in mixed coordinates. A more modern approach uses *Lie*-transforms, defined as *Hamiltonian* flows and therefore manifestly symplectic.

Let $H$ be a $C^{\infty}$-function on the cotangent bundle, i.e. $H:T^{\star}M \rightarrow \mathbf{R}$. On the cotangent bundle lives a natural symplectic form $\omega$. Given $x \in T^{\star}M$, there exist local coordinates $(q,p), q \in M, p \in T_q^{\star}M$, such that $\omega$ can be written as $\omega = \sum_{i=1}^{n} dq_i dp_i$. The coordinates $q$ and $p$ are called position and momentum. We say that the *Hamiltonian* system has $n$ degrees of freedom, where $n$ is the dimension of $M$.

This symplectic form can be used to define a vector field $X$ induced by the *Hamiltonian* as follows:

$$\iota_X \omega = dH,$$

where $\iota$ is the usual inner product of a vector with a two-form. In local coordinates we find, if we denote $X$ by $\sum_{i=1}^{n} (X_q^i \frac{\partial}{\partial q_i} + X_p^i \frac{\partial}{\partial p_i})$:

$$\begin{aligned} \iota_X \omega &= \iota_{\sum_{k=1}^{n}(X_q^k \frac{\partial}{\partial q_k} + X_p^k \frac{\partial}{\partial p_k})} \sum_{l=1}^{n} dq_l dp_l \\ &= \sum_{k=1}^{n} \sum_{l=1}^{n} (X_q^k dp_l - X_p^k dq_l) \delta_{kl} \\ &= \sum_{k=1}^{n} X_q^k dp_k - X_p^k dq_k. \end{aligned}$$

Since

$$dH = \sum_{k=1}^{n} \frac{\partial H}{\partial q_k} dq_k + \frac{\partial H}{\partial p_k} dp_k$$

we find that

$$X = \sum_{k=1}^{n} \frac{\partial H}{\partial p_k} \frac{\partial}{\partial q_k} - \frac{\partial H}{\partial q_k} \frac{\partial}{\partial p_k}.$$

This gives the familiar *Hamilton* equations:

$$\begin{aligned} \dot{q}_k &= \frac{\partial H}{\partial p_k}, \\ \dot{p}_k &= - \frac{\partial H}{\partial q_k}. \end{aligned}$$

Suppose now that $H$ has a critical point (0,0), i.e. a stationary point of the differential equation; Taking local coordinates around this critical point, $H$ can be written as follows. We let $H(0,0)$ be zero, since its value does not matter:

$$H(q,p) = H_2(q,p) + H_3(q,p) + \cdots,$$

where $H_n(q,p)$ is a homogeneous polynomial of degree $n$ . To make quantitatively explicit that we expand in a neighborhood of (0,0) we scale by $\epsilon$ : $q = \epsilon q^*, p = \epsilon p^*$. This results in

$$H^* = \epsilon^2 H_2(q^*,p^*) + \epsilon^3 H_3(q^*,p^*) + \cdots,$$

$$\omega^* = \epsilon^2 \sum_{k=1}^{n} dq^*{}_k dp^*{}_k.$$

Dividing both $\omega^*$ and $H^*$ by $\epsilon^2$ does not change the differential equation and we obtain a new *Hamiltonian* (where we drop the $\star$'s)

$$H = H_2(q,p) + \epsilon H_3(q,p) + \cdots.$$

This leads to differential equations of the type

$$\dot{x} = f^o(x) + \epsilon f^1(x) + \cdots$$

and we could define a normal form as before. This has the disadvantage that one cannot be sure that such a normal form is again *Hamiltonian*, since for this to happen the normalizing transformation has to be at least approximately symplectic, i.e. it must leave $\omega$ invariant.
Symplectic transformations can be generated as follows: Define a transformation $(p,q) \mapsto (\mathbf{p},\mathbf{q}) \in T^*\mathbf{R}$ implicitly by

$$\mathbf{q} = u(q,\mathbf{p}),$$

$$p = w(q,\mathbf{p}).$$

Then

$$dqdp = dq(\frac{\partial w}{\partial q}dq + \frac{\partial w}{\partial \mathbf{p}}d\mathbf{p}) = \frac{\partial w}{\partial \mathbf{p}}dqd\mathbf{p},$$

$$d\mathbf{q}d\mathbf{p} = (\frac{\partial u}{\partial q}dq + \frac{\partial u}{\partial \mathbf{p}})d\mathbf{p} = \frac{\partial u}{\partial q}dqd\mathbf{p}.$$

For this transformation to be symplectic one must have

$$\frac{\partial w}{\partial \mathbf{p}} = \frac{\partial u}{\partial q}.$$

We can solve this by taking

$$w = \frac{\partial K}{\partial q}, \qquad u = \frac{\partial K}{\partial \mathbf{p}},$$

where $K(q,\mathbf{p})$ is a function, usually called the generating function, since it generates the transformation. This can readily be generalized to $n$ degrees of freedom:

$$\mathbf{q}_k = \frac{\partial K}{\partial \mathbf{p}_k},$$

$$p_k = \frac{\partial K}{\partial q_k}$$

and the proof runs as follows

$$\begin{aligned}\omega &= \sum_{k=1}^{n} dq_k dp_k = \sum_{k=1}^{n} dq_k (\sum_{l=1}^{n} \frac{\partial^2 K}{\partial q_l \partial q_k} dq_l + \frac{\partial^2 K}{\partial \mathbf{p}_l \partial q_k} d\mathbf{p}_l) \\ &= \sum_{k=1}^{n}\sum_{l=1}^{n} \frac{\partial^2 K}{\partial q_l \partial q_k} dq_k dq_l + \sum_{k=1}^{n}\sum_{l=1}^{n} \frac{\partial^2 K}{\partial \mathbf{p}_l \partial q_k} dq_k d\mathbf{p}_l \\ &= \sum_{k=1}^{n}\sum_{l=1}^{n} \frac{\partial^2 K}{\partial \mathbf{p}_l \partial q_k} dq_k d\mathbf{p}_l.\end{aligned}$$

On the other hand we can compute

$$\omega = \sum_{k=1}^{n} d\mathbf{q}_k d\mathbf{p}_k = \sum_{k=1}^{n}\sum_{l=1}^{n} \frac{\partial^2 K}{\partial \mathbf{p}_l \partial q_k} dq_k d\mathbf{p}_l$$

and it follows that indeed $\omega = \omega$. To put this in the context of perturbation theory, we expand $K$ as we have expanded the Hamiltonian in a neighborhood of (0,0) :

$$K = K_2 + \epsilon K_3 + \epsilon^2 K_4 + \cdots .$$

Since we do not want to change $H_2$ by the transformation, $K_2$ has to generate the identity, and so

$$K_2(q,\mathbf{p}) = q\mathbf{p} = \sum_{k=1}^{n} q_k \mathbf{p}_k .$$

Then $(q,p) \mapsto (\mathbf{q},\mathbf{p})$ is a near-identity transformation. To find $K_3$ and $K_4$, Poisson-brackets enable us to do the computation in an efficient way. The Poisson-brackets { , } act on pairs of functions on the cotangent bundle and produce a new function. They can be defined by

$$\{G,H\} = \omega(X,Y),$$

where $\iota_X\omega = dH$ and $\iota_Y\omega = dG$. In local coordinates this implies

$$\begin{aligned}\{G,H\} = \omega(X,Y) &= -\sum_{i=1}^{n} dq_i dp_i (\sum_{k=1}^{n} Y_q^k \frac{\partial}{\partial q_k} + Y_p^k \frac{\partial}{\partial p_k}, \sum_{l=1}^{n} X_q^l \frac{\partial}{\partial q_l} + X_p^l \frac{\partial}{\partial p_l}) \\ &= \sum_{i=1}^{n}\sum_{k=1}^{n}\sum_{l=1}^{n} \delta_{ik}\delta_{il}(Y_q^k X_p^l - Y_p^k X_q^l) \\ &= \sum_{i=1}^{n} (\frac{\partial G}{\partial q_i}\frac{\partial H}{\partial p_i} - \frac{\partial G}{\partial p_i}\frac{\partial H}{\partial q_i}).\end{aligned}$$

We write

$$K = q\mathbf{p} + \epsilon K^\star .$$

Then

$$\mathbf{q} = \frac{\partial K}{\partial \mathbf{p}} = q + \epsilon \frac{\partial K^\star}{\partial \mathbf{p}},$$

$$p = \frac{\partial K}{\partial q} = \mathbf{p} + \epsilon \frac{\partial K^\star}{\partial q}$$

and

$$q = \mathbf{q} - \epsilon \frac{\partial K^\star}{\partial \mathbf{p}}(q,\mathbf{p}) = \mathbf{q} - \epsilon \frac{\partial K^\star}{\partial \mathbf{p}}(\mathbf{q},\mathbf{p}) + \epsilon^2 \frac{\partial^2 K^\star}{\partial q \partial \mathbf{p}}(\mathbf{q},\mathbf{p}) \frac{\partial K^\star}{\partial \mathbf{p}},$$

$$p = \mathbf{p} + \epsilon \frac{\partial K^\star}{\partial q}(q,\mathbf{p}) = \mathbf{p} + \epsilon \frac{\partial K^\star}{\partial q}(\mathbf{q},\mathbf{p}) - \epsilon^2 \frac{\partial^2 K^\star}{\partial q^2}(\mathbf{q},\mathbf{p}) \frac{\partial K^\star}{\partial \mathbf{p}}.$$

This implies

$$H(q,p) = H(\mathbf{q},\mathbf{p}) + \epsilon(\frac{\partial H}{\partial p}\frac{\partial K^\star}{\partial q}(\mathbf{q},\mathbf{p}) - \frac{\partial H}{\partial q}\frac{\partial K^\star}{\partial \mathbf{p}}(\mathbf{q},\mathbf{p}))$$
$$+ \epsilon^2(\frac{\partial H}{\partial q}\frac{\partial^2 K^\star}{\partial q \partial \mathbf{p}}\frac{\partial K^\star}{\partial \mathbf{p}} - \frac{\partial H}{\partial p}\frac{\partial^2 K^\star}{\partial q^2}\frac{\partial K^\star}{\partial \mathbf{p}} + ½\frac{\partial K^\star}{\partial \mathbf{p}}\frac{\partial^2 H}{\partial q^2}\frac{\partial K^\star}{\partial \mathbf{p}}$$
$$- \frac{\partial K^\star}{\partial \mathbf{p}}\frac{\partial^2 H}{\partial q \partial p}\frac{\partial K^\star}{\partial q} + ½\frac{\partial K^\star}{\partial q}\frac{\partial^2 H}{\partial p^2}\frac{\partial K^\star}{\partial q})$$
$$= H(\mathbf{q},\mathbf{p}) + \epsilon\{K^\star,H\} + ½\epsilon^2(\{K^\star,\{K^\star,H\}\} + \{H,\frac{\partial K^\star}{\partial q}\frac{\partial K^\star}{\partial \mathbf{p}}\}) + \cdots .$$

One should keep in mind that $H$ and $K$ are functions of $\epsilon$ , so this expression has to be rewritten, by substituting

$$H = H_2 + \epsilon H_3 + \epsilon^2 H_4 + \cdots ,$$
$$K^\star = K_3 + \epsilon K_4 + \cdots .$$

We find

$$\overline{H} = H_2 + \epsilon H_3 + \epsilon^2 H_4 + \epsilon\{K_3 + \epsilon K_4, H_2 + \epsilon H_3\}$$
$$+ ½\epsilon^2(\{K_3,\{K_3,H_2\}\} + \{H_2,\frac{\partial K_3}{\partial q}\frac{\partial K_3}{\partial \mathbf{p}}\})$$
$$= H_2 + \epsilon(H_3 + \{K_3,H_2\})$$
$$+ \epsilon^2(H_4 + \{K_3,H_3\} + \{K_4,H_2\}$$
$$+ ½\{K_3,\{K_3,H_2\}\} + ½\{H_2,\frac{\partial K_3}{\partial q}\frac{\partial K_3}{\partial \mathbf{p}}\}).$$

If we put $\overline{H}_3 = H_3 - \{H_2,K_3\}$ , this reads

$$\overline{H} = H_2 + \epsilon\overline{H}_3 + \epsilon^2(H_4 + \{K_3,\overline{H}_3\} - \{K_3,\{K_3,H_2\}\} + \{K_4,H_2\}$$
$$+ ½\{K_3,\{K_3,H_2\}\} + ½\{H_2,\frac{\partial K_3}{\partial q}\frac{\partial K_3}{\partial \mathbf{p}}\})$$
$$= H_2 + \epsilon\overline{H}_3 + \epsilon^2(H_4 + ½\{K_3,H_3 + \overline{H}_3\}$$
$$+ ½\{H_2,\frac{\partial K_3}{\partial q}\frac{\partial K_3}{\partial \mathbf{p}}\} + \{H_2,K_4\}).$$

We did not specify how we made the choice of $H_3$. Following the definition of normal forms of a vector field, it seems natural to require that $\overline{H}_3$ be in the kernel of $adH_2$, working on functions defined on the cotangent bundle, modulo constant functions, where

$$adH_2 : G \mapsto \{H_2, G\},$$

i.e. $H_2$ and $H_3$ must commute under *Poisson*-brackets. We work here modulo constants since adding a constant to a *Hamiltonian* does not change the *Hamilton* equations. This requires the same kind of splitting of the space of *Hamiltonians* in the kernel and the image of $adH_2$ as we met in our study of vector fields (with constant frequency ).

If $H_2 = p_1$ , the equation to be solved is simply

$$\frac{\partial K_3}{\partial q_1} = H_3 - \overline{H}_3$$

(since $\frac{\partial H_2}{\partial q_1} = 0$) and, if $q_1$ is an angular variable, we can take $\overline{H}_3$ to be the average of $H_3$ over $q_1$, i.e.

$$H_3(q_2, \cdots, q_n, p_1, \cdots, p_n) = \int_{S^1} H_3(q_1, \cdots, q_n, p_1, \cdots p_n) dq_1$$

(where, as usual, the measure is weighted such that $\int_{S^1} dq_1 = 1$) and

$$K_3(q_1, \cdots, q_n, p_1, \cdots, p_n) =$$

$$\int^{q_1} H_3(\phi, q_2, \cdots, q_n, p_1, \cdots, p_n) - \overline{H}_3(q_2, \cdots, q_n, p_1, \cdots, p_n) d\phi.$$

We may take $K_3$ with zero $q_1$-average. When $H_2 = G(p_1)$ we find, putting $\Omega = G'$,

$$\Omega(p_1) \frac{\partial K_3}{\partial q_1} = H_3 - \overline{H}_3.$$

Again, let $\overline{H}_3$ be the $q_1$-average of $H_3$ and

$$K_3(q,p) = \frac{1}{\Omega(p_1)} \int^{q_1} H_3(\phi, \hat{q}, p) - \overline{H}_3(\hat{q}, p) d\phi,$$

where $\hat{q} = (q_2, \cdots, q_n)$. This is only possible when $\Omega(p_1)$ is bounded away from zero uniformly, otherwise the whole asymptotic construction would fail near the singularity of $\Omega^{-1}$.

## 6.5. Normalization of Hamiltonians around equilibria

### 6.5.1. The generating function

The equilibria of a *Hamiltonian* vector field coincide with the critical points of the *Hamiltonian*. Suppose we have found such a critical point and consider it as the origin for local symplectic coordinates around the equilibrium. Since the value of the *Hamiltonian* at the critical point is not important, we take it to be zero, and we expand the *Hamiltonian* in the local coordinates in a *Taylor*-expansion:

$$H = H_2 + \epsilon H_3 + \epsilon^2 H_4 + \cdots,$$

where $H_k$ is homogeneous of degree k and $\epsilon$ is a scaling factor, related to the magnitude of the neighborhood that we take around the equilibrium. We shall assume $H_2$ to be in the following standard form

$$H_2 = \sum_{j=1}^{n} \frac{1}{2}\omega_j(q_j^2 + p_j^2)$$

(When some of the eigenvalues of $d^2H$ have equal magnitude, this standard form may not be obtainable. In two degrees of freedom this makes the 1:1- and the 1:-1-resonance exceptional (Cf. *Cushman* (Cus82a) and *van der Meer* (Mee82a)). We call $\omega = (\omega_1, \ldots, \omega_n)$ the *frequency-vector*. In the following discussion of resonances, it turns out that it is convenient to have two other coordinate systems at our disposal, i.e. *complex coordinates* and *action-angle variables*. We shall introduce these first. Let

$$x_j = q_j - ip_j,$$

$$y_j = q_j + ip_j.$$

Then

$$dx_j dy_j = 2idq_j dp_j.$$

In order to obtain the same vector field, and keeping in mind the definition

$$\iota_{X_H}\omega = dH,$$

we have to multiply the new Hamiltonian by $2i$ after the substitution of the new $x$ and $y$ coordinates in the old $H$ . Thus

$$H = H_2 + \epsilon H_3 + \cdots,$$

where

$$H_2 = i\sum_{j=1}^{n} \omega_j x_j y_j$$

( and $H_k$ is again homogeneous of degree k, this time in x and y).
The next transformation will be to action-angles variables. One should take care with this transformation since it is singular when a pair of coordinates vanishes. We put

$$x_j = (2\tau_j)^{\frac{1}{2}} e^{i\phi_j}, \qquad \phi_j \in S^1,$$

$$y_j = (2\tau_j)^{\frac{1}{2}} e^{-i\phi_j}, \qquad \tau_j \in (0,\infty).$$

Then

$$dx_j dy_j = 2id\phi_j d\tau_j.$$

Thus we have to divide the new Hamiltonian by the scaling factor $2i$ after substitution of the action-angle variables. We obtain

$$H_2 = \sum_{j=1}^{n} \omega_j \tau_j$$

and

$$H_l = \sum_{\|m\|_1 \leq l} h_l^m(\tau) e^{i<m,\phi>},$$

where $<m,\phi> = \sum_{j=1}^{n} m_j \phi_j$ , $m_j \in \mathbf{Z}$ and $\|m\|_1 = \sum_{j=1}^{n} |m_j|$ . The $h_l^m$ are homogeneous of degree $\frac{l}{2}$ in $\tau$ . Applying the same transformation to the generating function of the normalizing transformation K, we can write

$$K = K_2 + \epsilon K_3 + \epsilon^2 K^4 + \cdots,$$

where

$$K_l = \sum_{\|m\|_1 \leq l} k_l^m(\tau) e^{i<m,\phi>}.$$

The normal form equation is

$$\{H_2, K_3\} = H_3 - \overline{H}_3 \ , \ \{\overline{H}_3, H_2\} = 0,$$

$$\{K_3, H_2\} = \sum_{\|m\|_1 \leq 3} <m,\omega> k_3^m(\tau) e^{i<m,\phi>}.$$

We can solve the normal form equation:

$$k_3^m = \begin{cases} \dfrac{1}{<m,\omega>} h_3^m(\tau) & <m,\omega> \neq 0 \\ 0 & <m,\omega> = 0. \end{cases}$$

Then

$$\overline{H}_3 = \sum_{\substack{<m,\omega> = 0 \\ \|m\|_1 \leq 3}} h_3^m(\tau) e^{i<m,\phi>}$$

and $\overline{H}_3$ commutes with $H_2$ , i.e. $\{H_2, H_3\} = 0$. For $<m,\omega>$ nonzero, but very small, this introduces large terms in the asymptotic expansion *(small divisors)*. In that case it might be better to treat $<m,\omega>$ as zero, and split of the part of $H_2$ that gives exactly zero, and consider this as the unperturbed problem. Suppose $<m,\omega> = \delta$ and $<m,\omega^o> = 0$, where $\omega$ and $\omega^o$ are close, then

$$H_2 = \sum \omega_j \tau_j = \sum \omega_j^o \tau_j + \sum (\omega_j - \omega_j^o) \tau_j.$$

We say that $m \in \mathbf{Z}^n$ is an *annihilator* of $\omega^o$ if

$$<m,\omega^o> = 0.$$

For the annihilator we use again the norm $\|m\|_1 = \sum_{j=1}^{n} |m_j|$ .

**6.5.1.1. Definition**

If the annihilators with norm less than or equal to $M + 2$, span a *codim 1* sublattice of $\mathbf{Z}^n$, and $M \in \mathbf{N}$ is minimal, then we say that $\omega^o$ defines a *genuine Mth order resonance.* ($M$ is minimal means that $M$ represents the lowest natural number corresponding with genuine resonance).

Examples of genuine resonances are: for $n=2$ : $k:l$, with $m^1=(-l,k)$ and $k+l>2$. For $n=3$ : 1:2:1, with $m^1=(2,-1,0)$ and $m^2=(0,-1,2)$, 1:2:2, with $m^1=(2,-1,0)$ and $m^2=(2,0,-1)$, 1:2:3, with $m^1=(2,-1,0)$, $m^2=(1,1,-1)$, 1:2:4, $\cdots$, with $m^1=(2,-1,0)$ and $m^2=(0,2,-1)$.

## 6.5.2. Some remarks on genuine resonances

Our definitions imply the following

**6.5.2.1. Lemma**

When $\omega$ is a genuine Mth order resonance, there exists a $\lambda \in \mathbf{R}$ such that $\lambda\omega \in \mathbf{Z}^n$.

Although this hardly merits a proof, we shall nevertheless spell out the details, if only to introduce the notation for the next proof.

**Proof**

Take $n-1$ annihilators spanning the sublattice orthogonal to $\omega$ over $\mathbf{Q}$ (not over $\mathbf{Z}$, since it may not always be possible to span the sublattice with only $n-1$ annihilators over the integers, take eg $\omega = (1,1,1) \in \mathbf{Z}^3$). We label the annihilators $m^k$ and write down the equations:

$$\sum_{j=1}^{n} m_j^k \omega_j = 0, \qquad k = 1, \ldots, n-1.$$

We extend this with

$$\sum_{j=1}^{n} m_j^n \omega_j = \mu \neq 0$$

such that $\hat{M} = (m_j^i)_{i,j=1,\ldots,n}$ is an invertible matrix with integer coefficients. We denote by $M_i$ the $(n-1)\times(n-1)$ sub-matrix obtained by leaving out the $i^{th}$ column in $M$. Then, by *Cramer*'s rule, we find

$$\omega_i = \mu \frac{\det(M_i)}{\det(\hat{M})} \in \mu\mathbf{Q}.$$

Take $\lambda = \dfrac{\det(\hat{M})}{\mu}$ and we have $\lambda\omega_i \in \mathbf{Z}$ ,or $\lambda\omega \in \mathbf{Z}^n$. Thus we can always rescale the problem such that $\omega \in \mathbf{Z}^n$ and we can take $\omega$ minimal in the sense that

$$gcd(\omega_1, \ldots, \omega_n) = 1. \qquad \square$$

This allows us to extend the system with $\lambda = 1$ (Theorem 3 in *Hecke*

(Hec81a)) and we find

$$\omega_i = \frac{\det(M_i)}{\det(\hat{M})}.$$

Let $\|\omega\|_\infty = \max_{j=1,\ldots,n} |\omega_j|$. Then

$$\|\omega\|_\infty \leqslant (M+1)^{n-1}.$$

To show this we formulate the following

**6.5.2.2. Lemma**

Let $\tilde{M}$ be a $n \times (n-1)$-matrix with integer coefficients, such that

$$\max_{i=1,\ldots,n-1} \|m_i\|_1 \leqslant M + 2,$$

where $m^i = (m_j^i)_{j=1,\ldots,n}$ and $\tilde{M} = (m_j^i)_{j=1,\ldots,n}^{i=1,\ldots,n-1}$ Then there exists, if rank $\tilde{M} = n-1$ , a unique $\omega \neq 0$, $\omega \in \mathbf{Z}^n$ such that $\tilde{M}\omega = 0$ and $gcd(\omega_1, \ldots, \omega_n) = 1$ with

$$\|\omega\|_\infty \leqslant (M+1)^{n-1}.$$

**Proof (*P. Noordzij, F. van Schagen*, 1982, private communication)**

We have

$$|\omega_i| \leqslant |\det(M_i)|$$

since $\det(\hat{M}) \in \mathbf{Z}$ and $\det(\hat{M}) \neq 0$. ($\hat{M}$ is the extension of $\tilde{M}$ with $<m_n,\omega> = 1$).

Since the $i$th column of $\tilde{M}$ must have a nonzero element, say $m_i^l$, due to the fact that the rank of K is $n-1$, there is a row in $M_i$ such that

$$\sum_{\substack{j=1 \\ j \neq i}}^{n} |m_j^l| \leqslant M+1,$$

implying

$$|\det M_i| \leqslant \sum_{\substack{j=1 \\ j \neq i}}^{n} |m_j^l| \, |\det(M_j^l)| \leqslant (M+1) \max_j |\det(M_j^l)|,$$

where $M_j^l$ is obtained by expanding $M_i$ with respect to $m_j^l$ (cofactor). Since individual elements of $M_i$ are less than (or equal to) $M+1$ in absolute value, we can repeat this argument with smaller and smaller sub-determinants to obtain

$$|\det(M_i)| \leqslant (M+1)^{n-1}$$

(The power is most easily checked on low-dimensional problems) and hence

$$\|\omega\|_\infty \leqslant (M+1)^{n-1}.$$

□

By uniqueness, the same estimate holds for a Mth order genuine resonance in normalized form, ie in $\mathbf{Z}^n$ and relatively prime components. A easy corollary is that the number of genuine Mth order resonances is bounded. Of more importance is the fact that this estimate does help to restrict the number of possibilities when one computes all Mth order resonances. This computation is trivial in two degrees of freedom, but in three degrees of freedom there is no method but to test all possible $\omega$. A list of genuine first order resonances appeared in (Aa,79a) and of second order resonances in (Ver83a).

The obtained estimate is moreover sharp: consider

$$\omega = (1, n-1, (n-1)^2, \ldots, (n-1)^{n-1}).$$

## 6.6. Normal form polynomials

In the sequel we shall not carry out all details of normalizing concrete *Hamiltonians*, but we shall assume that the *Hamiltonian* is already in normal form. The idea is to study the general normal form, and determine its properties depending on the free parameters. These parameters can be computed in concrete problems by the normalization procedure. We shall first determine which polynomials are in normal form with respect to

$$H_2 = \tfrac{1}{2}\sum_{j=1}^{n} \omega_j (q_j^2 + p_j^2).$$

Changing to complex coordinates, and introducing a multiindex notation, we can write a general polynomial term, derived from a real one, as

$$P_{kl} = i(Dx^k y^l + \bar{D}x^l y^k),$$

where $x^k = x_1^{k_1} \ldots x_n^{k_n}, y^l = y_1^{l_1} \ldots y_n^{l_n}$, and $k_i, l_i \geq 0, i = 1, \ldots, n$.

Since

$$H_2 = i\sum_{j=1}^{n} \omega_j x_j y_j$$

we find

$$\{H_2, P_{kl}\} = \sum_{j=1}^{n} \left(\frac{\partial H_2}{\partial x_j}\frac{\partial P_{kl}}{\partial y_j} - \frac{\partial P_{kl}}{\partial x_j}\frac{\partial H_2}{\partial y_j}\right)$$

$$= \sum_j \omega_j \left(x_j \frac{\partial}{\partial x_j} - y_j \frac{\partial}{\partial y_j}\right)(Dx^k y^l + \bar{D}x^l y^k)$$

$$= (Dx^k y^l + \bar{D}x^l y^k) < \omega, k-l > .$$

So $P_{kl} \in ker(adH_2)$ is equivalent with $<\omega, k - l> = 0$

This is of course nothing but the usual resonance relation. In action-angle variables one only looks at the difference $k - l$, and the homogeneity condition puts a bound on this difference. The most important resonance term arises for $\|k + l\|_1$ minimal. For two degrees of freedom, with $\omega_1$ and $\omega_2 \in \mathbf{N}$ and relatively prime, the resonance term is

$$P = Dx_1^{\omega_2} y_2^{\omega_1} + \overline{D} y_1^{\omega_2} x_2^{\omega_1}.$$

Terms with $k = l$ are also resonant. They are called self-interaction terms and are polynomials in the variables $x_i y_i$ or $\tau_i$ .

# 7. Hamiltonian Systems

## 7.1. Introduction

In *Hamiltonian* mechanics, the small parameter necessary to do asymptotics is usually obtained by localizing the system around some well-known solution, e.g. an equilibrium. As we shall see, the part played by the small parameter in the normal form of the *Hamiltonian* determines the asymptotic estimates which we can obtain. In the various resonance cases which we shall discuss, these estimates take different forms the theory of which is based on the preceding chapters with special extensions for the *Hamiltonian* context.

There are a few global results that one has to bear in mind while doing such a local analysis. A readable introduction to these qualitative aspects can be found in *Berry* (Ber78a). A useful reference source, also including the theoretical physics literature is *Helleman* (Hel80a). Following *Poincaré*, the main interest has been in determining the 'skeleton' of the flow, ie the low-dimensional invariant manifolds; more specifically, one searches for equilibria, periodic orbits and invariant tori.
Equilibria constitute in general no problem, since they can be found by solving a set of ordinary equations. Although this may be a far from trivial task in practice, we shall always consider it done.
To obtain periodic orbits is another matter. A basic theorem is due to *Lyapounov* (Lya47a) for analytic *Hamiltonians* with $n$ degrees of freedom: if the eigenfrequencies of the linearized *Hamiltonian* are independent over **Z** near a stable equilibrium point, there exist $n$ families of periodic solutions filling up smooth 2-dimensional manifolds going through the equilibrium point. Fixing the energy level near an equilibrium point, one finds from

these families $n$ periodic solutions, usually called the normal modes of the system. These solutions can be considered as a continuation of the $n$ families of periodic solutions that one finds for the linearized equations of motion.
The assumption of non-resonance (the eigenfrequencies independent over $\mathbf{Z}$) has been dropped in a basic theorem by *Weinstein* (Wei73a): for an $n$ degrees of freedom *Hamiltonian* system near a stable equilibrium, there exist (at least) $n$ periodic solutions for fixed energy. Some of these solutions may be a continuation of linear normal mode families but others are clearly not related to these, as we shall see (in fact, in certain resonance cases the linear normal modes cannot continue).
Since the paper by *Weinstein* several results appeared in which all these periodic solutions have been indiscriminately referred to as normal modes. We shall not adopt this confusing terminology; a normal mode will be a periodic solution 'restricted' to a two-dimensional invariant subspace of the linearized system.
The existence of invariant tori around the periodic solutions is predicted by the *Kolmogorov*, *Arnol'd*, *Moser*- theorem (or *KAM*-theorem) which is a collection of statements the first proofs of which have been provided by *Arnol'd* and *Moser*; see(Arn83a) and(Mos71a) and further references there. According to this theorem, under rather general assumptions, the energy manifold ($(2n-1)$-dimensional in a $n$ degrees of freedom system) contains an infinite number of $n$-dimensional tori, invariant under the flow. In a neighborhood of an equilibrium point in phase space, most of the orbits are located on these tori, or somewhat more precise: as $\epsilon \to 0$, the measure of orbits between the tori tends to zero. We find only regular behavior by the asymptotic approximations, and this can be interpreted as a further quantitative specification of the *KAM*-theorem. For instance, if we describe phase space by regular orbits with error of $O(\epsilon^k)$ on the time-scale $\epsilon^{-m}$ $(k,m>0)$ we have clearly an upper bound on how wild solutions can be on this time-scale. It may improve our insight into the possible richness of the phase-flow to enumerate some dimensions:

| | | | |
|---|---|---|---|
| degrees of freedom: | 2 | 3 | n |
| dimension of phase-space: | 4 | 6 | 2n |
| dimension of energy manifold: | 3 | 5 | 2n-1 |
| dimension invariant tori: | 2 | 3 | n |

The tori are described by keeping the $n$ actions fixed and varying the $n$ angles, which makes them $n$-dimensional. The tori are imbedded in the energy manifold and there is clearly no escape possible from between the tori if $n=2$. For $n \geqslant 3$, orbits can get out between tori; For this process, called *Arnol'd*-diffusion (Arn64a), *Nekhoroshev*(Nek77a) showed that it takes place on at least an exponential time-scale of the order $\frac{1}{\epsilon} e^{\epsilon^{-m}}$ where $\epsilon^2$ is a measure for the energy with respect to an equilibrium position.

## 7.2. Resonance zone

We have seen that *Hamiltonian* systems with $n$ degrees of freedom have (at least) $n$ families of periodic orbits in the neighborhood of a stable equilibrium, by *Weinstein*'s 1973 theorem. This is the minimal number of periodic orbits for fixed energy, due to the resonance the actual number may be higher. The general normal form of an *Hamiltonian* near equilibrium depends on parameters, and the dimension of the parameter space depends on $n$, on the resonance, and on the order of the normal form.
If we fix all these, we can define the *zone of resonance* to be those values of the frequencies for which the normal form has more than $n$ short (i.e. $O(1)$)-periodic orbits for a given energy level. The boundary of the resonance zone is contained in the *bifurcation set* of the resonance. For practical and theoretical reasons one is often interested in the dependency of the bifurcation set on the energy.

## 7.3. Canonical variables adapted to the resonance

In the normal form of the *Hamiltonian* near equilibrium, one has combination angles $<m,\phi>$. It will often simplify the problem to take these angles as new variables, eliminating in the process one (fast) combination. For our notation we refer to chapter 6. There we found that for a given resonant $\omega$ there exists a matrix $\hat{M} \in GL(n,\mathbf{Z})$ such that

$$\hat{M}\omega = \begin{pmatrix} 0 \\ 0 \\ \cdot \\ \cdot \\ 0 \\ 1 \end{pmatrix} + o(1),$$

where the o(1)-term represents detuning. We drop this term to keep the notation simple. Let

$$\psi = \hat{M}\phi.$$

Then $\dot{\psi}_i = 0 + o(1)$ , $i = 1, \cdots, n-1$ and $\dot{\psi}_n = 1 + o(1)$. The action variables are found by the dual definition:

$$\tau = \hat{M}^{\dagger} r$$

(where $\hat{M}^{\dagger}$ denotes the transpose of $\hat{M}$). This defines a symplectic change of coordinates from $(\phi,\tau)$ to $(\psi,r)$ variables, since

$$\sum_i d\phi_i d\tau_i = \sum_{ij} \hat{M}_{ji} d\phi_i dr_j = \sum_j d\psi_j dr_j .$$

We shall often denote $r_n$ with E. E is the only variable independent of the extension from $M$ to $\hat{M}$.

## 7.4. Two degrees of freedom

### 7.4.1. Introduction

A two-degrees of freedom system is characterized by a four-dimensional phase flow and to visualize this flow is already complicated. To obtain a geometric picture is useful for the full understanding of a dynamical system, even if one is only interested in the asymptotics. It also helps to have a clear picture in mind of the linearized flow. We give a list of possible ways of looking at a certain problem. To be specific, we assume that the *Hamiltonian* of the linearized system is positive definite. The indefinite case is difficult from the asymptotic point of view, since solutions can grow without bounds, although one can still use the results on periodic orbits.

a) The *Poincaré*-mapping: Fixing the energy we have, near a stable equilibrium, flow on a compact manifold, diffeomorphic to $S^3$. As the energy manifold is compact, we have a priori bounds for the solutions. One should note that this sphere need not be a sphere in the (strict) geometric sense.

   One can take a plane, locally transversal to the flow on the energy manifold; this plane is mapped into itself under the flow, which defines a *Poincaré*-mapping. This map is easier to visualize, since it is two-dimensional. Note that, due to the local character of the map, the *Poincaré* map does not necessarily describe all orbits; in a situation with two normal modes for example, the map around one will not describe the other.

b) Projection into 'physical space': In the older literature one finds often a representation of the solutions by projection onto the base (or configuration) space, with coordinates $q_1$, $q_2$. In physical problems that is the space where one can see things happen. Under this projection periodic orbits typically look like algebraic curves, the order determined by the resonance. If they are stable and surrounded by tori, these tori project as tubes around these algebraic curves.

c) A visual representation that is also useful in systems with more than one degree of freedom is to plot the actions $\tau_i$ as functions of time; some authors prefer to use the amplitudes $(2\tau_i)^{1/2}$ instead.

d) As we shall see, in two degrees of freedom systems only one slowly varying combination angle $\psi_1$ plays a part. It is possible then, to plot the actions as functions of $\psi_1$; in this representation the periodic solutions show up as critical points of the $\tau,\psi_1$-flow.

e) The picture of the periodic solution and their stability changes as the parameters of the *Hamiltonian* change. To illustrate these changes we draw bifurcation diagrams which illustrate the existence and stability of solutions and also the branching off and vanishing of periodic orbits. These bifurcation diagrams take various forms, see for instance § 7.4.3 and § 7.5.4.

It is useful to have various means of illustration at our disposal as the complications of higher dimensional phase flow are not so easy to grasp from only one type of illustration. It may also be useful for the reader to consult the pictures in (Abr78a) and (Abr83a).

Some remarks finally on the linearized flow of a two degrees of freedom system. The *Hamiltonian* is

$$H = \omega_1 \tau_1 + \omega_2 \tau_2$$

and the equations of motion

$$\dot{\phi}_i = \omega_i \ , \ \dot{\tau}_i = 0 \ , \ i = 1,2,$$

corresponding with harmonic solutions

$$\begin{bmatrix} q \\ p \end{bmatrix} = \begin{bmatrix} (2\tau_1(0))^{1/2} \sin(\phi_1(0) + \omega_1 t) \\ (2\tau_2(0))^{1/2} \sin(\phi_2(0) + \omega_2 t) \\ (2\tau_1(0))^{1/2} \cos(\phi_1(0) + \omega_1 t) \\ (2\tau_2(0))^{1/2} \cos(\phi_2(0) + \omega_2 t) \end{bmatrix}.$$

If $\dfrac{\omega_1}{\omega_2} \notin \mathbf{Q}$, there are two periodic solutions for each value of the energy, the normal modes $\tau_1 = 0$ and $\tau_2 = 0$. If $\dfrac{\omega_1}{\omega_2} \in \mathbf{Q}$, all solutions are periodic.

We fix the energy, choosing a positive constant $E_o$ with

$$\omega_1 \tau_1 + \omega_2 \tau_2 = E_o.$$

In $q,p$-space this represents an ellipsoid which we identify with $S^3$. The energy manifold is invariant under the flow but also, in this linear case, $\tau_1$ and $\tau_2$ are conserved quantities corresponding with invariant manifolds in $S^3$. What do these invariant manifolds look like?

They are described by two equations

$$\omega_1 \tau_1 + \omega_2 \tau_2 = E_o$$

and

$$\omega_1 \tau_1 = E_1 \ \textit{or} \ \omega_2 \tau_2 = E_2.$$

$E_1$ and $E_2$ are both positive and their sum equals $E_o$. Choosing $E_1 = 0$ corresponds with a normal mode in the $\tau_2$-component (all energy $E_o$ in the second degree of freedom); as we know from harmonic solutions this is a circle lying in the $q_2,p_2$-plane. The same reasoning applies to the case $E_2 = 0$ with a normal mode in the $\tau_1$-component. Consider one of these circles lying in $S^3$. The other circle passes through the center of the first one, because the center of the circle corresponds to a point where one of the actions $\tau$ is zero, which makes the other action maximal, and thus part of a normal mode. We have the following figure

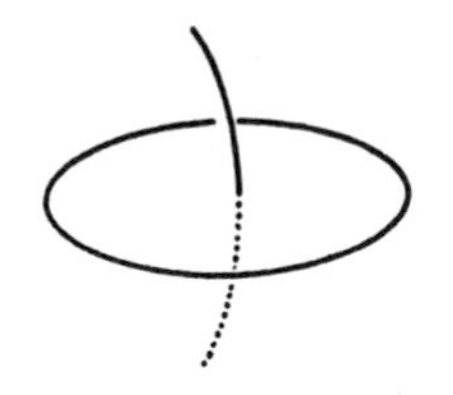

**Figure 7.4.1-1**

On the other hand, if we draw the second circle first, the picture must be the same, be it in another plane. This leads to

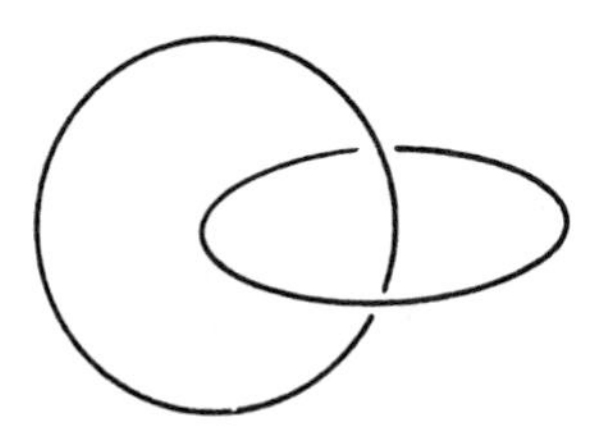

**Figure 7.4.1.-2**

The normal modes are linked and they form the extreme cases of the invariant manifolds $\omega_i \tau_i = E_i$ , $i = 1,2$. What do these manifolds look like when $E_1 E_2 > 0$? A *Poincaré*-mapping is easy to construct, it looks like a family of circles in the plane.

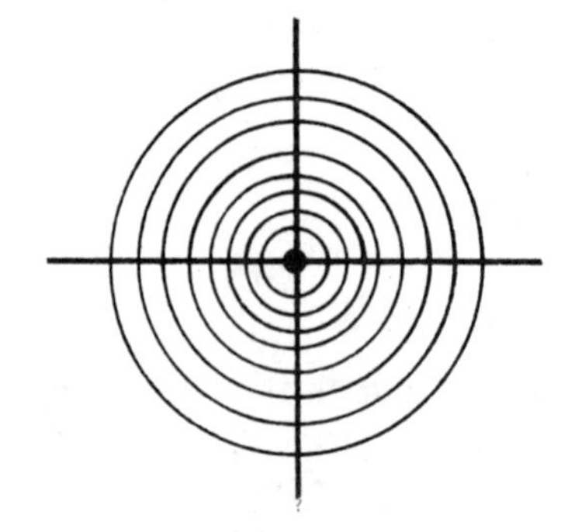

**Figure 7.4.1-3**

The center is a fixed point of the mapping corresponding with the normal mode in the second degree of freedom. The boundary represents the normal mode in the first degree of freedom; note that the normal mode does not belong to the *Poincaré*-mapping as the flow is not transversal here to the $q_1,p_1$-plane. Starting on one of the circles in the $q_1,p_1$-plane with $E_1,E_2 > 0$, the return map will produce another point on the circle. If

$\frac{\omega_1}{\omega_2} \notin \mathbf{Q}$ the starting point will never be reached again; If $\frac{\omega_1}{\omega_2} \in \mathbf{Q}$ the orbits close after some time, i.e. the starting point will be attained. These orbits are called periodic orbits in *general position.* Clearly the invariant manifolds $\omega_1\tau_1 + \omega_2\tau_2 = E_o$, $\omega_1\tau_1 = E_1$ are invariant tori around the normal modes.

We could have concluded this immediately, but with less detail of the geometric picture near the normal modes, by considering the action-angle variables $\tau,\phi$ and their equations of motion: if the $\tau_i$ are fixed, the $\phi_i$ are left to be varying and they describe the manifold we are looking for, the torus $\mathbf{T}^2$. Thus the energy surface has the following *foliation:* there are two invariant, linked circles, the normal modes and around these invariant tori filling up the sphere.

### 7.4.2. The $k:l$-resonance, $0<k<l$, $(k,l)=1$

A good reference for the theory of periodic solutions for systems in resonance is (Dui84a).

In complex coordinates, the normal form of the $k:l$-resonance is (Cf. chapter 6 on normal forms)

$$H = i(\omega_1 x_1 y_1 + \omega_2 x_2 y_2) + i\epsilon^{k+l-2}(D x_1^l y_2^k + \overline{D} y_1^l x_2^k)$$
$$+ i\epsilon^2\left(\frac{A}{2}(x_1 y_1)^2 + B(x_1 y_1)(x_2 y_2) + \frac{C}{2}(x_2 y_2)^2\right) + \cdots,$$

where $A,B,C \in \mathbf{R}$ and $D \in \mathbf{C}$ . Putting $D = |D| e^{i\alpha}$, we have in action-angle coordinates,

$$H = \omega_1\tau_1 + \omega_2\tau_2 + \epsilon^{k+l-2} |D| (2\tau_1)^{\frac{l}{2}} (2\tau_2)^{\frac{k}{2}} \cos(l\phi_1 - k\phi_2 + \alpha)$$
$$+ \epsilon^2(A\tau_1^2 + 2B\tau_1\tau_2 + C\tau_2^2) + \cdots .$$

In the sequel we shall drop the dots. Let $\delta = l\omega_1 - k\omega_2$ be the (small) detuning parameter. The resonance matrix can be taken as

$$\hat{M} = \begin{bmatrix} l & -k \\ k^* & l^* \end{bmatrix} \in SL(2,\mathbf{Z}).$$

Following § 7.3 we introduce adapted resonance coordinates:

$$\psi_1 = l\phi_1 - k\phi_2 + \alpha,$$
$$\psi_2 = k^*\phi_1 + l^*\phi_2,$$
$$\tau_1 = lr + k^*E,$$
$$\tau_2 = -kr + l^*E.$$

In the normal form given above, of the angles only the combination denoted by $\psi_1$ plays a part; we shall therefore replace $\psi_1$ by $\psi$ in this section on two degrees of freedom systems. Then we have

$$H = (\omega_1 k^* + \omega_2 l^*)E + \delta r$$

$$+\epsilon^{k+l-2}|D|(2lr+2k^*E)^{\frac{l}{2}}(-2kr+2l^*E)^{\frac{k}{2}}\cos\psi$$

$$+\epsilon^2((Al^2-2Bkl+Ck^2)r^2+2(Alk^*+B(ll^*-kk^*)-Ckl^*)Er$$

$$+(Ak^{*^2}+2Bk^*l^*+Cl^{*^2})E^2).$$

The angle $\psi_2$ is not present in the *Hamiltonian* so that E is an integral of the equations of motion induced by the normal form ( $\dot{E} = -\dfrac{\partial H}{\partial \psi_2} = 0$ ). As $E$ corresponds with the $H_2$ part of the *Hamiltonian*, the quantity $E$ is conserved for the full problem to $O(\epsilon+\delta)$ for all time. Let

$$\Delta_1 = \begin{vmatrix} A & k \\ B & l \end{vmatrix}, \qquad \Delta_2 = \begin{vmatrix} B & k \\ C & l \end{vmatrix},$$

$$\Delta_1^* = \begin{vmatrix} A & k^* \\ B & -l^* \end{vmatrix}, \qquad \Delta_2^* = \begin{vmatrix} B & k^* \\ C & -l^* \end{vmatrix}.$$

Then

$$H = (\omega_1 k^* + \omega_2 l^*)E + \delta r + \epsilon^{k+l-2}|D|(2lr+2k^*E)^{\frac{l}{2}}(-2kr+2l^*E)^{\frac{k}{2}}\cos\psi$$

$$+\epsilon^2((l\Delta_1 - k\Delta_2)r^2 + 2(k^*\Delta_1 + l^*\Delta_2)Er + (k^*\Delta_1^* + l^*\Delta_2^*)E^2).$$

This leads to the differential equation

$$\dot{r} = \epsilon^{k+l-2}|D|(2k^*E + 2lr)^{\frac{l}{2}}(2l^*E - 2kr)^{\frac{k}{2}}\sin\psi,$$

$$\dot{\psi} = \delta + \epsilon^{k+l-2}|D|2(2k^*E+2lr)^{\frac{l}{2}-1}(2l^*E-2kr)^{\frac{k}{2}-1}$$

$$\times((l^2l^* - k^2k^*)E - kl(k+l)r)\cos\psi$$

$$+ 2\epsilon^2((l\Delta_1 - k\Delta_2)r + (k^*\Delta_1 + l^*\Delta_2)E).$$

To complete the system we have to write down $\dot{E}=0$ and the equation for $\psi_2$. We shall omit these equations in the sequel. In the beginning of this section we characterized the normal modes by putting one of the actions equal to zero. For periodic orbits in general position we have $\tau_i \neq 0$ and constant, $i=1,2$. This implies that a periodic solution in general position corresponds with $r$ constant during the motion, resulting in the condition

$$\sin(\psi) = 0 \;\; i.e. \;\; \psi = 0, \pi$$

during periodic motion. So we also have to look for stationary points of the equation for $\psi$ and this implies that $\delta$ must be of $O(\epsilon)$ if $k+l=3$ or of $O(\epsilon^2)$ if $k+l \geq 4$.

### 7.4.3. The 1:2-resonance

The 1:2-resonance is the only first order resonance in two degrees of freedom systems. A convenient choice for $\hat{M}$ turns out to be

$$\hat{M} = \begin{pmatrix} 2 & -1 \\ 1 & 0 \end{pmatrix}$$

to find the differential equations

$$\dot{r} = \epsilon|D|(-2r)^{½}(2E+4r)\sin\psi,$$

$$\dot{\psi} = \delta+2\epsilon|D|(-2r)^{-½}(-E-6r)\cos\psi + O(\epsilon^2).$$

Periodic solutions are obtained by finding the stationary points of this equation, and that leads to $\sin\psi=0$ and $\cos\psi=\pm1$. Moreover,

$$-r\delta^2 = 2\epsilon^2|D|^2(E+6r)^2.$$

The normal modes are given by $\tau_1=0$ and $\tau_2=0$
Since

$$\tau_1 = 2r + E,$$

$$\tau_2 = -r.$$

this corresponds to $r=-\dfrac{E}{2}$ and $r=0$. The second one can never be a solution, but the first relation leads to

$$\frac{E}{2}\delta^2 = 8\epsilon^2|D|^2E^2$$

or

$$\delta = \pm4\epsilon|D|E^{½} \qquad \textit{(Bifurcation set).}$$

The domain of resonance is defined by the inequality

$$|\delta|<4\epsilon|D|E^{½}.$$

Strictly speaking, we cannot draw any conclusions about the existence of normal mode solutions from our analysis, since the action-angle variables are singular at the normal modes. Therefore, we shall now analyze this situation somewhat more rigorously, using *Morse* theory. We let

$$x_j = q_j - ip_j,$$

$$y_j = q_j + ip_j.$$

The normal form is then, in these real coordinates

$$H=H_2+\epsilon|D|(\cos\alpha((q_1^2-p_1^2)q_2+2q_1p_1p_2)-\sin\alpha((q_1^2-p_1^2)p_2-2q_1p_1q_2)).$$

We want to study the normal mode given by the equation $q_1=p_1=0$. We put

$$p_2 = -(2\tau)^{½}\sin\phi,$$

$$q_2 = (2\tau)^{1/2}\cos\phi.$$

This symplectic transformation induces the new *Hamiltonian*

$$H = \tfrac{1}{2}\omega_1(q_1^2+p_1^2)+\omega_2\tau+\epsilon|D|(2\tau)^{1/2}(\cos(\phi-\alpha)(q_1^2-p_1^2)-2\sin(\phi-\alpha)q_1p_1).$$

We shall take as our unperturbed problem the *Hamiltonian* $H^o$ , defined as

$$H^o = \tfrac{1}{2}(q_1^2 + p_1^2) + 2\tau.$$

To analyze the *normal mode,* we consider a periodic orbit as a critical orbit of $H$ with respect to $H^o$ . To show that the normal mode is indeed a critical orbit and to compute its stability type, we use *Lagrange* multipliers: The extended *Hamiltonian* $\mathbf{H}$ is defined as

$$\mathbf{H} = \mu H^o + H, \qquad \mu\in\mathbf{R},$$

where we fix the energy level by $H^o = E\in\mathbf{R}$ . We find $\mu$ from

$$d\mathbf{H} = 0$$

and substituting $q_1 = p_1 = 0$.
Indeed,

$$d\mathbf{H}=\begin{pmatrix}(\mu+\omega_1)q_1\\(\mu+\omega_1)p_1\\0\\2\mu+\omega_2\end{pmatrix}+2\epsilon|D|(2\tau)^{1/2}\begin{pmatrix}\cos(\phi-\alpha)q_1-\sin(\phi-\alpha)p_1\\-\cos(\phi-\alpha)p_1-\sin(\phi-\alpha)q_1\\O(q_1^2+p_1^2)\\O(q_1^2+p_1^2)\end{pmatrix}.$$

The critical orbit is given by the vector $(0,0,\phi,\frac{E}{2})$ and its tangent space is spanned by $(0,0,1,0)$ . The kernel of $dH^o$ is spanned by $(1,0,0,0)$, $(0,1,0,0)$ and $(0,0,1,0)$ and this implies that the normal bundle of the critical orbit is spanned by $(1,0,0,0)$ and $(0,1,0,0)$ . The second derivative of $\mathbf{H}$ , $d^2\mathbf{H}$ , is defined on this normal bundle and can easily be computed:

$$d^2\mathbf{H} = (\mu + \omega_1)\begin{pmatrix}1&0\\0&1\end{pmatrix} + 2\epsilon|D|(2\tau)^{1/2}\begin{pmatrix}\cos(\phi-\alpha)&-\sin(\phi-\alpha)\\-\sin(\phi-\alpha)&-\cos(\phi-\alpha)\end{pmatrix}.$$

It follows from $d\mathbf{H} = 0$ that $\mu = -\frac{\omega_2}{2}$, so

$$d^2\mathbf{H}=\begin{pmatrix}\frac{\delta}{2}+2\epsilon E^{1/2}|D|\cos(\phi-\alpha)&-2\epsilon E^{1/2}|D|\sin(\phi-\alpha)\\-2\epsilon E^{1/2}|D|\sin(\phi-\alpha)&\frac{\delta}{2}-2\epsilon E^{1/2}|D|\cos(\phi-\alpha)\end{pmatrix}.$$

We find

$$Tr(d^2\mathbf{H}) = \delta,$$

$$Det(d^2\mathbf{H}) = \frac{\delta^2}{4} - 4\epsilon^2E|D|^2.$$

When $Det(d^2\mathbf{H})>0$, the normal mode is elliptic since $d^2\mathbf{H}$ is a definite form; when $Det(d^2\mathbf{H})<0$, the normal mode is hyperbolic. The bifurcation value is

$$\delta = \pm 4\epsilon E^{1/2} |D|,$$

as we found before.
We can now give the picture of the periodic orbits versus detuning:

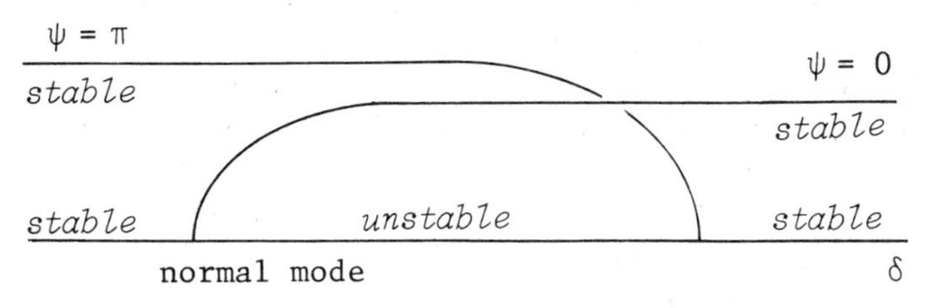


**Figure 7.4.3-1**

The crossing of the two elliptic orbits is not a bifurcation, since the solutions are $\pi$ out of phase.
So there are two possibilities: either there are two elliptic solutions (the minimum configuration) or there are two elliptic orbits and one hyperbolic. The bifurcation of the elliptic from the normal mode does not violate any index argument, because it is a *flip-orbit:*

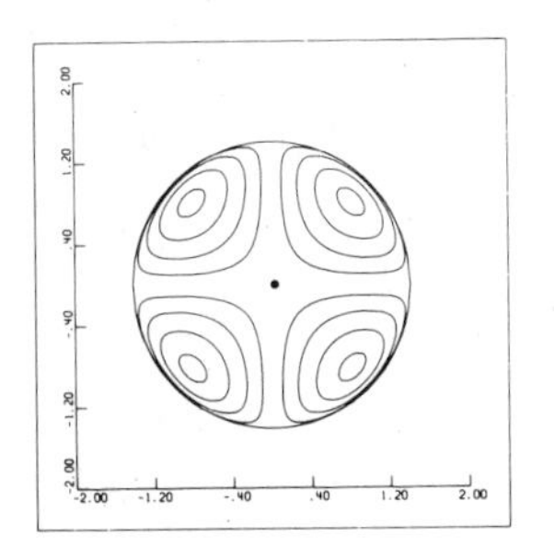


**Figure 7.4.3-2** *Poincaré section for the* 1:2*-resonance. The fixed point in the center corresponds with a hyperbolic normal mode; the four elliptic fixed points correspond with two stable periodic solutions.*

In a *Poincaré*-section it looks like two orbits, but these two invariant points should be identified.

As we have seen, the normalized *Hamiltonian* is of the form

$$\overline{H} = H_2 + \epsilon \overline{H}_3 + O(\epsilon^2)$$

in which $\overline{H}_3$ stands for the normalized cubic part. We have found that $H_2$ corresponds with an integral of the normalized system, accuracy $O(\epsilon)$ and validity for all time with respect to the orbits of the original system. Of course, the *Hamiltonian* $\overline{H}$ is itself an integral of the normalized system, and we can take as two independent, *Poisson*-commuting integrals $H_2$ and $\overline{H}_3$. This discussion generalizes to the $n$-degrees of freedom case, but with 2 degrees of freedom we found the maximal number of integrals, and are therefore able to conclude that the normalized system is always integrable. This simplifies the analysis of 2 degrees of freedom systems considerably.

We define the *momentum mapping* as follows:

$$M:T^*\mathbf{R}^2 \to \mathbf{R}^2,$$

$$M(q,p)=(H_2(q,p),\overline{H}_3(q,p)).$$

Using this mapping, we can analyze the foliation induced by the two integrals. In general, $M^{-1}(x,y)$ will be a torus, or empty. For special values the inverse image consists of a circle, which in this case is also a periodic orbit.

The asymptotic accuracy of this procedure follows from the estimates on the solutions of the differential equation. For the 1:2-resonance we have $O(\epsilon)$-accuracy on the time-scale $\frac{1}{\epsilon}$.

### 7.4.4. The symmetric 1:1-resonance

The differential equations for the general $k:l$-resonance, $0<k<l$, do not apply to the 1:1-resonance. We can, however, use them for what we shall call the *symmetric 1:1-resonance* (or the 2:2-resonance). The normal form for the 1:1-resonance is complicated, with many parameters. If we impose, however, mirror symmetry in each of the symplectic coordinates, that is invariance of the *Hamiltonian* under the four transformations:

$$M_1:(q_1,p_1,q_2,p_2)\mapsto(-q_1,p_1,q_2,p_2),$$

$$M_2:(q_1,p_1,q_2,p_2)\mapsto(q_1,-p_1,q_2,p_2),$$

$$M_3:(q_1,p_1,q_2,p_2)\mapsto(q_1,p_1,-q_2,p_2),$$

$$M_4:(q_1,p_1,q_2,p_2)\mapsto(q_1,p_1,q_2,-p_2),$$

then this simplifies the normal form considerably. Since this symmetry assumption is natural in several applications (where one must realize that the assumption need not be valid for the original *Hamiltonian*, only for the normal form), one could even say that the symmetric 1:1-resonance is more important than the general 1:1-resonance; at least it merits a separate treatment.

Note that two normal modes exist in this case; we leave this to the reader. We take

$$\hat{M} = \begin{pmatrix} l & -k \\ k^* & l^* \end{pmatrix} = \begin{pmatrix} 2 & -2 \\ 2 & 2 \end{pmatrix}.$$

The differential equations are

$$\dot{r} = 16\epsilon^2 |D| (E^2 - r^2)\sin\psi,$$

$$\dot{\psi} = \delta + 2\epsilon^2 |D| (-16r)\cos\psi + 4\epsilon^2((\Delta_1 - \Delta_2)r + (\Delta_1 + \Delta_2)E).$$

The stationary solutions are determined by

$$\sin\psi = 0 \quad \Rightarrow \quad \cos\psi = \pm 1,$$

$$\delta \pm 32\epsilon^2 |D| r + 4\epsilon^2(\Delta_1 - \Delta_2)r + 4\epsilon^2(\Delta_1 + \Delta_2)E = 0.$$

Let

$$\delta = 4\epsilon^2 E\Delta,$$

$$r = Ex.$$

Then

$$\Delta \pm 8|D| x_\pm + (\Delta_1 - \Delta_2)x_\pm + (\Delta_1 + \Delta_2) = 0$$

or

$$x_\pm = -\frac{\Delta + \Delta_1 + \Delta_2}{\Delta_1 - \Delta_2 \pm 8|D|}$$

if $|\Delta_1 - \Delta_2| \neq 8|D|$.
Since, by definition, for orbits in general position

$$\tau_1 = 2(E + r) > 0,$$

$$\tau_2 = 2(E - r) > 0,$$

we have $|x| < 1$ and this gives the following bifurcation equations:

$$\Delta = -(\Delta_1 + \Delta_2),$$

$$\Delta = -2(\Delta_1 \pm 4|D|).$$

The linearized equation at the stationary point $(\psi_0, r_0)$ is

$$\dot{r} = 16\epsilon^2 |D| (E^2 - r_0{}^2)\cos(\psi_0)\psi,$$

$$\dot{\psi} = -32\epsilon^2 |D| \cos(\psi_0) r + 4\epsilon^2(\Delta_1 - \Delta_2)r,$$

where $\cos\psi_0 = \pm 1$ .
The eigenvalues are given by

$$\lambda^2 - 16\epsilon^2 |D| (E^2 - r_0{}^2)\cos\psi_0(-32\epsilon^2 |D| \cos\psi_0 + 4\epsilon^2(\Delta_1 - \Delta_2)) = 0.$$

The orbit is elliptic if

$$8|D| > \pm(\Delta_1 - \Delta_2)$$

and hyperbolic otherwise, if not on the bifurcation. The bifurcation takes place when

$$8|D| = |\Delta_1 - \Delta_2|.$$

This is a so called global bifurcation; for this value of the parameters both normal modes bifurcate at the same moment, the equation for the stationary points is degenerate and in general one has to go to higher order approximations to see what happens. Despite its degenerate character, this global bifurcation keeps turning up in applications, cf. *Verhulst* (Ver79a) and *Sanders* (San78b).

### 7.4.5. The 1:3-resonance

#### 7.4.5.1. Periodic orbits in general position

There are two second order resonances in two degrees of freedom systems: the 1:1- and the 1:3-resonance. In this section we shall treat the 1:3. We take

$$\hat{M} = \begin{bmatrix} 3 & -1 \\ 1 & 0 \end{bmatrix}.$$

The differential equations are

$$\dot{r} = \epsilon^2 \,|\, D \,|\, (2E+6r)^{\frac{3}{2}} (-2r)^{1/2} \sin\psi,$$

$$\dot{\psi} = \delta + 2\epsilon^2 \,|\, D \,|\, (2E+6r)^{1/2} (-2r)^{-1/2} (-E-12r)\cos\psi + 2\epsilon^2((3\Delta_1 - \Delta_2)r + \Delta_1 E).$$

This leads to the following equation for the stationary points

$$\sin\psi = 0,$$

$$(\delta + 2\epsilon^2((3\Delta_1 - \Delta_2)r + \Delta_1 E))^2(-2r) = 4\epsilon^4 \,|\, D \,|^2 (2E+6r)(-E-12r)^2.$$

This equation is cubic in $r$, so in principle we can find an explicit solution. Let

$$r = Ex,$$

$$\delta = \Delta\epsilon^2 E.$$

Then

$$(\Delta + 2\Delta_1 + 2(3\Delta_1 - \Delta_2)x)^2(-2x) = 8 \,|\, D \,|^2 (1+3x)(1+12x)^2.$$

Let

$$\alpha = \Delta + 2\Delta_1,$$

$$\beta = 2(3\Delta_1 - \Delta_2),$$

$$\gamma = 2 \,|\, D \,|.$$

Then we have

$$-(\alpha + \beta x)^2 x = \gamma^2(1 + 27x + 216x^2 + 432x^3)$$

or

$$(432\gamma^2 + \beta^2)x^3 + (216\gamma^2 + 2\alpha\beta)x^2 + (27\gamma^2 + \alpha^2)x + \gamma^2 = 0.$$

We shall not give the explicit solutions, but we determine the bifurcation set of this equation. First we transform to the standard form for cubic equations:

$$y^3 + uy + v = 0.$$

Let

$$ax^3 + bx^2 + cx + d = 0$$

and put

$$y = x - \frac{b}{3a}.$$

Then we obtain

$$u = \frac{(ac - \frac{b^2}{3})}{a^2},$$

$$v = \frac{(da^2 - \frac{1}{3}abc + \frac{2}{27}b^3)}{a^3}.$$

The bifurcation set of this standard form is the well known cusp:

$$27v^2 + 4u^3 = 0$$

and, after some extensive calculations, we find this to be equivalent to a homogeneous polynomial of degree 12 in $\alpha, \beta$ and $\gamma$ . After factoring out, the bifurcation equation can be written as:

$$\alpha^4 + 54\alpha^2\gamma^2 - 243\gamma^4 - \frac{1}{3}\alpha^3\beta - 27\alpha\beta\gamma^2 + \frac{9}{4}\beta^2\gamma^2 = 0$$

(We neglect here the isolated bifurcation plane $12\alpha = \beta$). Let us consider the curve $P=0$, with

$$P(\alpha,\beta,\gamma) = \alpha^4 + 54\alpha^2\gamma^2 - 243\gamma^4 - \frac{1}{3}\alpha^3\beta - 27\alpha\beta\gamma^2 + \frac{9}{4}\beta^2\gamma^2 =$$

$$- \frac{1}{3}(\alpha^2 + 27\gamma^2)^2 + \frac{1}{3}\alpha^3(4\alpha - \beta) + \frac{9}{4}(4\alpha - \beta)(8\alpha - \beta)\gamma^2.$$

This suggests the following transformation:

$$X = \alpha,$$

$$Y = (27)^{½}\gamma,$$

$$Z = ½(4\alpha - \beta).$$

Then the induced P is

$$P^* = -\frac{1}{3}(X^2 + Y^2 - XZ)^2 + \frac{1}{3}(X^2 + Y^2)Z^2.$$

Putting $Z = 1$, we have the equation for the cardioid:

$$(X^2 + Y^2 - X)^2 = (X^2 + Y^2).$$

Changing to polar coordinates

$$X = r\cos\theta,$$

$$Y = r\sin\theta,$$

this takes the form

$$r = 1 + \cos\theta.$$

Another representation is the following: by intersecting the curve with the pencil of circles

$$X^2 + Y^2 - 2X = tY$$

we obtain

$$(X + tY)^2 = (2X + tY)$$

and this implies

$$(t^2 - 1)Y + 2tX = 0.$$

Substituting this in the equation for the circle bundle, we find

$$X = \frac{2(1 - t^2)}{(1 + t^2)^2}$$

and

$$Y = \frac{4t}{(1 + t^2)^2},$$

so we have a rational parametrization of the bifurcation curve.

**7.4.5.2. Normal mode**

The normal form of the 1:3-resonance in real coordinates $q$ and $p$ is given by

$$\begin{aligned} H = {} & \tfrac{1}{2}(\omega_1(q_1^2 + p_1^2) + \omega_2(q_2^2 + p_2^2)) + \\ & \tfrac{1}{2}|D|\epsilon^2((\cos\alpha + i\sin\alpha)(q_1 - ip_1)^3(q_2 + ip_2) \\ & + (\cos\alpha - i\sin\alpha)(q_1 + ip_1)^3(q_2 - ip_2) \\ & + \frac{\epsilon^2}{4}(A(q_1^2 + p_1^2)^2 + 2B(q_1^2 + p_1^2)(q_2^2 + p_2^2) + C(q_2^2 + p_2^2)^2) \\ = {} & \tfrac{1}{2}(\omega_1(q_1^2 + p_1^2) + \omega_2(q_2^2 + p_2^2)) \\ & + \epsilon^2|D|(\cos\alpha((q_1^2 - 3p_1^2)q_1q_2 + (3q_1^2 - p_1^2)p_1p_2 \\ & - \sin\alpha((q_1^3 - 3p_1^2q_1)p_2 - (3q_1^2p_1 - p_1^3)q_2) \\ & + \tfrac{1}{4}\epsilon^2(A(q_1^2 + p_1^2)^2 + 2B(q_1^2 + p_1^2)(q_2^2 + p_2^2) + C(q_2^2 + p_2^2)^2). \end{aligned}$$

We study the normal mode $q_1 = p_1 = 0$, so we put

$$p_2 = -(2\tau)^{1/2}\sin\phi,$$

$$q_2 = (2\tau)^{1/2}\cos\phi$$

and we find

$$H = \frac{\omega_1}{2}(q_1^2 + p_1^2) + \omega_2 \tau$$

$$+ \epsilon^2 |D| (2\tau)^{1/2} (\cos(\phi - \alpha)(q_1^2 - 3p_1^2) q_1 - (3q_1^2 - p_1^2) p_1 \sin(\phi - \alpha))$$

$$+ \epsilon^2 (\frac{A}{4}(q_1^2 + p_1^2)^2 + B\tau(q_1^2 + p_1^2) + C\tau^2).$$

Let $\mathbf{H} = \mu H^o + H$, $\quad H^o = 3E$ , where $H^o = \frac{1}{2}(q_1^2 + p_1^2) + 3\tau$. Then

$$d\mathbf{H} = \begin{bmatrix} (\mu+\omega_1)q_1 \\ (\mu+\omega_1)p_1 \\ 0 \\ \omega_2+3\mu \end{bmatrix} + \epsilon^2 \begin{bmatrix} O(q_1^2+p_1^2) \\ O(q_1^2+p_1^2) \\ O(q_1^2+p_1^2) \\ O(q_1^2+p_1^2) \end{bmatrix} + \epsilon^2 \begin{bmatrix} 2B\tau q_1 \\ 2B\tau p_1 \\ 0 \\ 2C\tau + O(q_1^2+p_1^2) \end{bmatrix}$$

and

$$d^2\mathbf{H} = \begin{bmatrix} \mu + \omega_1 + 2B\tau\epsilon^2 & 0 \\ 0 & \mu + \omega_1 + 2B\tau\epsilon^2 \end{bmatrix}.$$

Since $\omega_2 + 3\mu + 2C\epsilon^2\tau = 9$ and $3\tau = E$ ,

$$d^2\mathbf{H} = \begin{bmatrix} -\frac{\omega_2}{3} + \omega_1 + 2(B - \frac{C}{3})E\epsilon^2 & 0 \\ 0 & -\frac{\omega_2}{3} + \omega_1 + 2(B - \frac{C}{3})E\epsilon^2 \end{bmatrix}$$

$$= \frac{1}{3}\begin{bmatrix} \delta + 2(3B - C)E\epsilon^2 & 0 \\ 0 & \delta + 2(3B - C)E\epsilon^2 \end{bmatrix}.$$

Unless $\delta + 2(B - \frac{C}{3})E\epsilon^2 = 0$ , this is a definite form, and the normal mode is elliptic. The bifurcation value, where $d^2\mathbf{H} = 0$ , marks the 'flipping through' of a hyperbolic orbit, in such a way that this orbit changes its phase with a factor $\pi$ in the *Poincaré* section transversal to the normal mode.

### 7.4.6. Higher order resonances

#### 7.4.6.1. Introduction

We speak of a *higher order resonance* when $k + l > 4$. The differential equations in normal form have solutions characterized by two different time scales; they are generated, as we shall see, by $\epsilon$-terms of order 2 and of order $k + l - 2$. This structure of the equations enables us to treat the higher order resonances all at once and we shall find the periodic orbits without making assumptions on $k$ and $l$. We present the calculation first, the general idea will be discussed in § 7.4.6.3.

### 7.4.6.2. Periodic orbits in general position

The equations are

$$\dot{r} = \epsilon^{k+l-2} | D | (2k^* E + 2lr)^{\frac{l}{2}} (2l^* E - 2kr)^{\frac{k}{2}} \sin\psi,$$

$$\dot{\psi} = \delta + 2\epsilon^2((l\Delta_1 - k\Delta_2)r + (k^*\Delta_1 + l^*\Delta_2)E) + O(k+l-2) + O(\epsilon^4).$$

To find periodic solutions we put $\sin\psi = 0$ and we find

$$\delta + 2\epsilon^2((l\Delta_1 - k\Delta_2)r + (k^*\Delta_1 + l^*\Delta_2)E = O(\epsilon^{k+l-2}) + O(\epsilon^4).$$

This equation does not necessarily have a solution. Let us scale

$$r = Ex,$$

$$\delta = 2E\epsilon^2\Delta.$$

Then we have to solve

$$\Delta + k^*\Delta_1 + l^*\Delta_2 + (l\Delta_1 - k\Delta_2)x = 0,$$

with $-\dfrac{k^*}{l} < x < \dfrac{l^*}{k}$ ( where $\dfrac{l^*}{k} - (-\dfrac{k^*}{l}) = \dfrac{1}{lk} > 0$ since $\hat{M} \in SL(2,\mathbf{Z})$ ).
Since

$$x = -\frac{(\Delta + k^*\Delta_1 + l^*\Delta_2)}{l\Delta_1 - k\Delta_2}, \qquad l\Delta_1 - k\Delta_2 \neq 0,$$

we have the condition on the parameters

$$-\frac{k^*}{l} < -\frac{(\Delta + k^*\Delta_1 + l^*\Delta_2)}{l\Delta_1 - k\Delta_2} < \frac{l^*}{k}.$$

This implies that the width of the resonance zone is given by

$$2\epsilon^2 E \left| \frac{\Delta_1}{k} - \frac{\Delta_2}{l} \right|$$

(the parameter, in terms of which the resonance zone has been defined is $\Delta$, the detuning).

In this domain we have two periodic orbits, and it is easy to see that one is elliptic, the other hyperbolic. This conclusion does not hold near the normal modes.

### 7.4.6.3. Asymptotic estimates

In this section we discuss the asymptotics of higher order resonances; for details of the proofs see (San78a). The equations which we used in the preceding analysis are of the form

$$\dot{r} = \epsilon^{k+l-2} f(r) \sin(\psi) + \cdots,$$

$$\dot{\psi} = \epsilon^2 g(r) + \epsilon^{k+l-2} h(r) + \cdots;$$

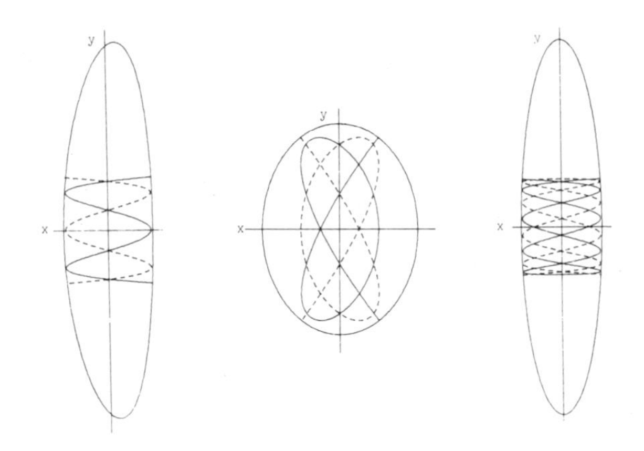


**Figure 7.4.6.2-1** *Projections into base space for the resonances 4:1, 4:3 and 9:2; cf. § 7.4.1, option b. The stable (full line) and unstable (- - -) periodic solutions are lying in the resonance manifold. The closed boundary is the curve of zero-velocity.*

$f(r)$, $g(r)$ and $h(r)$ are abbreviations for the expressions from the previous section; this system has to be supplemented by equations for $E$ and $\psi_2$.

The right hand side of the equations starts with terms of $O(\epsilon^2)$ and, using the theory of chapter 3, it is easy to obtain estimates like

$$r(t)=r(0)+O(\epsilon) \text{ on the time-scale } \frac{1}{\epsilon^2}$$

This means that on this time-scale no appreciable change of the variable $r$ takes place. To improve our insight in higher order resonance we note that the right hand side of the equation for $r$ is $O(\epsilon^{k+l-2})$ with $k+l-2 \geqslant 3$ and of the equation for $\psi$ is $O(\epsilon^2)$. In the spirit of chapter 5 we can consider $\psi$ to be rapidly varying with respect to the variable $r$ and it is then natural to average the system over the angle $\psi$.

This procedure breaks down where $\psi$ is not rapidly varying, i.e. in the domain where $g(r)$ is equal or near to zero. Note that the equation $g(r)=0$, characterizes the orbits in general position found in § 7.9.1. The equation $g(r)=0$ defines the (so-called) resonance manifold $M$ in the phase space.
The two domains in phase space which we need for the asymptotic estimates can now be introduced as follows.
$D_I$, the neighborhood of the resonance manifold $M$; this is the *inner* boundary layer: introducing the distance $d(p,M)$ for a point $p$ in 4-space to the manifold $M$ we have

$$D_I=\{p \mid d(p,M)=O(\epsilon^{\frac{k-l-4}{3}})$$

$D_O$, the remaining part of 4-space, the *outer* domain.

The outer domain $D_O$ will be the part of phase space where, to a certain approximation, there is no exchange of energy between the two degrees of freedom. The flow can be described as a simple, nonlinear continuation of the linearized flow on a long time-scale. This is expressed in terms of asymptotic estimates as follows:

**7.4.6.3.1. Theorem**

Consider the equations for $r,\psi$ and $E$ with initial conditions in the outer domain $D_O$ and the initial value problem

$$\dot{\tilde{\psi}}=2\epsilon^2 E\Delta+2\epsilon^2[(l\Delta_1-k\Delta_2)r+(k^*\Delta_1+l^*\Delta_2)E]\ ,\ \tilde{\psi}(0)=\psi(0)$$

then we have the estimates

$$r(t)-r(0), E(t)-E(0), \psi(t)-\tilde{\psi}(t)=O(\epsilon^{\frac{k+l-4}{6}})$$

on the time-scale $\epsilon^{-\frac{k+l}{2}}$.

## 7.5. Three degrees of freedom

### 7.5.1. Introduction

In contrast with the case of two degrees of freedom systems, the literature on this subject is still restricted. One of the reasons is undoubtedly the enormous increase in complexity of the expressions with the number of degrees of freedom; in the case of three degrees of freedom $H_3$ contains 56 terms, $H_4$ 126 terms. It is a question of considerable practical interest how to handle such longer expressions analytically. We shall find that by the process of normalization it is possible to obtain a drastic reduction of the size of these expressions.

One might wonder: are there new theoretical questions in systems with more than two degrees of freedom, are the questions not merely extensions of the same problems in a more complicated setting ? To some extent this is true with respect to the analysis of periodic solutions of the normalized *Hamiltonian*. Note however that the question of stability of these solutions is more difficult. In the case of two degrees of freedom the critical points of the equations for $r$ and $\psi$ (§ 7.4.2) will be elliptic or hyperbolic, characteristics which follow from a linear analysis. The existence of two-dimensional tori around these periodic solutions and the corresponding approximate integrals of motion which are valid for all time, then guarantee rigorously stability in the case of elliptic critical points of the reduced system. This property of rigorous results of a combined invariant tori/quasi-linear analysis argument is lost in the case of three degrees of freedom. In this case we find again elliptic and hyperbolic orbits and there exist corresponding surrounding invariant tori around the elliptic orbits, but these are 3-dimensional in a 5-dimensional sphere, so the tori do not divide the sphere into pieces, as in the lower-dimensional case. An easy way to see this is to consider only the actions. One can identify a torus with constant action-variables with a point on a $(n-1)$-simplex, where $n$ is the number of degrees of freedom. For $n=2$ the point does divide the interval into two pieces, but for $n=3$ it does not divide the triangle into pieces. This topological fact gives rise to the so called *Arnol'd diffusion* (for a discussion, see (Lic83a)). In the sequel we shall call periodic solutions corresponding with elliptic orbits again stable; note however that now we have stability only in a formal sense.

Another fundamental difference can be described as follows. In systems with two degrees of freedom we always find two integrals of the normalized *Hamiltonian* providing us with a complete description of the phase flow. This is expressed by saying that the normalized *Hamiltonian* is integrable. In the case of three degrees of freedom we still have two integrals, but we need three for the system to be integrable. To find a third integral is a nontrivial problem: in some cases it can be shown to exist, but there are also cases where it has been shown that a third analytic integral does not exist (Dui84b). This makes the global description of the phase flow of the normalized system essentially more difficult in the case of three degrees of freedom.

Another open question is the asymptotic analysis of three degrees of freedom systems. In a number of cases, for instance the genuine first order resonances, the analytic difficulties can be overcome, and a complete analysis is possible of the periodic orbits and their stability. There are some results on second order resonances, but the higher order resonances have never been analyzed.

### 7.5.2. The order of resonance

In this section we only consider *Hamiltonians* near a stable equilibrium point and at exact resonance; we put

$$H_2=\sum_{i=1}^{3} \tfrac{1}{2}\omega_i(q_i^2+p_i^2)\,, \quad \omega_i \in \mathbf{N}\,, \quad i=1,2,3.$$

Following § 6.5, we consider $k \in \mathbf{Z}^3$ and $k$-vectors such that $\sum_{i=1}^{3} \omega_i k_i = 0$. We identify annihilation vectors $k$ and $k'$ if $k+k'=0$. The number $\kappa=\sum_{i=1}^{3} |k_i|$, the norm of $k$, determines to which order we have to normalize; however, to characterize the possible interactions between the three degrees of freedom on normalizing to $H_\kappa$, we need another quantity. Compare for example the resonances 1:2:3 and 1:2:5. On normalizing to $H_3$ ($\kappa=3$) we have for the 1:2:3-resonance the annihilating vectors $(2,-1,0)$ and $(1,1,-1)$; for the 1:2:5-resonance only $(2,-1,0)$. Up till $H_3$, or in the language of asymptotic approximations: up till an $O(\epsilon)$-approximation on the time-scale $\frac{1}{\epsilon}$, the 1:2:3-resonance displays full interaction between all three degrees of freedom, the 1:2:5-resonance decouples at this level to a two degrees of freedom system and a one degree of freedom system. The case of full interaction between all three degrees of freedom was called a *genuine resonance* in(Aa,79a). To indicate the number of annihilating vectors at a certain order $\kappa$ we introduce the interaction number $\sigma_\kappa$; intuitively, the larger $\sigma_\kappa$ is, the more complex the analysis will appear to be. There are however no mathematical theorems to confirm this intuition and to measure exactly the complexity of any system in resonance. The same article contains a list of genuine first order resonances, and we reproduce it in Table I, each resonance with its interaction number.

| *resonance* | $\sigma_3$ | $\sigma_4$ |
|---|---|---|
| *1 : 2 : 1* | 3 | *1* |
| *1 : 2 : 2* | 2 | *1* |
| *1 : 2 : 3* | 2 | 2 |
| *1 : 2 : 4* | 2 | *1* |

**Table I**

| resonance | $\sigma_3$ | $\sigma_4$ |
|---|---|---|
| 1 : 1 : 1 | 0 | 6 |
| 1 : 1 : 3 | 0 | 5 |
| 1 : 2 : 5 | 1 | 1 |
| 1 : 2 : 6 | 1 | 1 |
| 1 : 3 : 3 | 0 | 3 |
| 1 : 3 : 4 | 1 | 1 |
| 1 : 3 : 5 | 0 | 3 |
| 1 : 3 : 6 | 1 | 1 |
| 1 : 3 : 7 | 0 | 2 |
| 1 : 3 : 9 | 0 | 2 |
| 2 : 3 : 4 | 1 | 1 |
| 2 : 3 : 6 | 1 | 1 |

**Table II**

(The reader may verify for instance that for the 1:2:1- resonance annihilating $k$-vectors are $(2,-1,0)$, $(0,-1,2)$ and $(1,-1,1)$). For the sake of completeness we also list the 12 genuine second order resonances with their interaction numbers in Table II. The first two cases, 1:1:1 and 1:1:3, appear to be the most complicated, followed by the resonances 1:3:3 and 1:3:5. Of course, symmetry assumptions may change all this.

### 7.5.3. Periodic orbits and integrals

The quadratic part of the *Hamiltonian* is

$$H_2 = \sum_{i=1}^{3} \tfrac{1}{2}\omega_i(q_i^2 + p_i^2) \quad , \ \omega_i \in \mathbf{N} \ , \ i = 1,2,3.$$

Using action-angle variables $\tau,\phi$ this part of the *Hamiltonian* looks like

$$H_2 = \sum_{i=1}^{3} \omega_i \tau_i .$$

Normalizing $H_3$ we find at most two linearly independent combinations of the three angles $\phi_i$; we shall denote these combination angles by $\psi_1$ and $\psi_2$. As discussed earlier, $H_2$ will be an approximate integral of the system, an exact integral of the normal form. In phase-space $H_2 = constant$ corresponds with $S^5$.

Periodic solutions are found as critical points of the normalized *Hamiltonian* reduced to $H_2$. In practice this involves the elimination of one action, the energy, leaving us with two action and two angle variables. The critical points are thus characterized by four eigenvalues; a pair of conjugate imaginary eigenvalues will be denoted in our pictures by $E$ (elliptic), a pair of opposite real eigenvalues by $H$ (hyperbolic), and the degenerate situation with zero eigenvalues by $O$. In section 7.4.1 we discussed visual presentations of the phase flow and periodic solutions. In the case of three degrees of freedom the following visualization (suggested by *R.Cushman*) is useful. We forget the angular variables and only plot the actions. For given energy, the set of allowable action values is a 2-simplex (triangular domain).

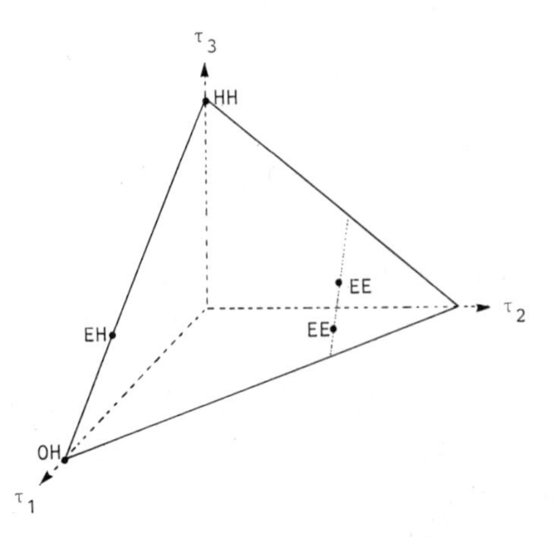


**Figure 7.5.3-1** *Action simplex; dots indicate periodic solutions, normal modes are at the vertices. The stability characteristics are denoted by E, H and O.*

The periodic solutions are points in this simplex, since they have fixed angles and actions. Note that according to (Wei73a) at least three periodic solutions exist for each energy value. To draw invariant surfaces might be impossible in this representation, unless the angular variables do not play a role. The normal modes are the vertices of the simplex. The linear stability is indicated by the two pairs of eigenvalues; for instance *EE* means two conjugate pairs of imaginary eigenvalues, *OH* means two eigenvalues zero and two real, *HH* means two real pairs, etc.

### 7.5.4. The 1:2:1-resonance

The first study of this relatively complicated case is by *van der Aa* and *Sanders* (Aa,79a), for an improved version (there were some errors in the calculations) see (Aa,83a). We shall always assume that the three degrees of freedom systems are in exact resonance, avoiding the analytical difficulties which seem to characterize the detuned problem.

In action-angle variables the normal form of $H_3$ is

$$\overline{H}=\tau_1+2\tau_2+\tau_3+2\epsilon(2\tau_2)^{1/2}[a_1\tau_1\cos(2\phi_1-\phi_2-a_2)$$
$$+a_3(\tau_1\tau_3)^{1/2}\cos(\phi_1-\phi_2+\phi_3-a_4)+a_5\tau_3\cos(2\phi_3-\phi_2-a_6)],$$

where $a_i, i=1,\ldots,6$ are real constants. Using the combination angles

$$2\psi_1=2\phi_1-\phi_2-a_2,$$
$$2\psi_2=2\phi_3-\phi_2-a_6,$$

we find the equations of motion (with $\eta=\frac{1}{2}a_2+\frac{1}{2}a_6-a_4$)

$$\dot{\tau}_1=2\epsilon(2\tau_2)^{1/2}[2a_1\tau_1\sin(2\psi_1)+a_3(\tau_1\tau_3)^{1/2}\sin(\psi_1+\psi_2+\eta)],$$
$$\dot{\tau}_2=-2\epsilon(2\tau_2)^{1/2}[2a_1\tau_1\sin(2\psi_1)+a_3(\tau_1\tau_3)^{1/2}\sin(\psi_1+\psi_2+\eta)+a_5\tau_3\sin(2\psi_2)],$$
$$\dot{\tau}_3=2\epsilon(2\tau_2)^{1/2}[a_3(\tau_1\tau_2)^{1/2}\sin(\psi_1+\psi_2+\eta)+2a_5\tau_3\sin(2\psi_2)],$$

$$\dot{\psi}_1=\epsilon(2\tau_2)^{1/2}[2a_1\cos(2\psi_1)+a_3(\frac{\tau_3}{\tau_1})^{1/2}\cos(\psi_1+\psi_2+\eta)]$$
$$-\frac{\epsilon}{(2\tau_2)^{1/2}}[a_1\tau_1\cos(2\psi_1)+a_3(\tau_1\tau_3)^{1/2}\cos(\psi_1+\psi_2+\eta)+a_5\tau_3\cos(2\psi_2)],$$

$$\dot{\psi}_2=\epsilon(2\tau_2)^{1/2}[a_3(\frac{\tau_1}{\tau_3})^{1/2}\cos(\psi_1+\psi_2+\eta)+2a_5\cos(2\psi_2)]$$
$$-\frac{\epsilon}{(2\tau_2)^{1/2}}[a_1\tau_1\cos(2\psi_1)+a_3(\tau_1\tau_3)^{1/2}\cos(\psi_1+\psi_2+\eta)+a_5\tau_3\cos(2\psi_2)].$$

As expected $H_2=\tau_1+2\tau_2+\tau_3$ is an integral of the normalized system. Analyzing the critical points of the equation of motion we find in the general case 7 periodic orbits (for each value of the energy) of the following three types:

1) one unstable normal mode in the $\tau_2$-direction.
2) two stable periodic solutions in the $\tau_2=0$ hyperplane.
3) two stable and two unstable periodic solutions in general position (i.e. $\tau_1\tau_2\tau_3>0$).

For some time now one has been looking for a third integral of the normalized system. *Van der Aa* (Aa,83a) obtained some negative results; *Duistermaat* (Dui84b) (see also (Dui84c) ) reconsidered the problem as follows. At first the normal form given above is simplified again by using the action of a certain linear symplectic transformation leaving $H_2$ invariant;

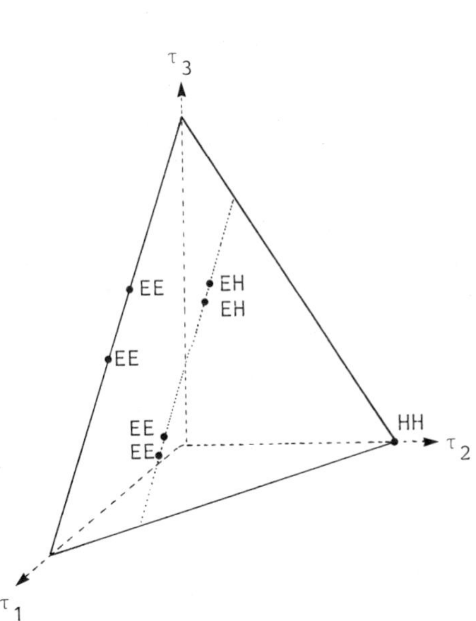


**Figure 7.5.4-1** *Action simplex for the the* 1:2:1*-resonance.*
this removes two parameters from the normal form. Then one observes that all solutions of the *Hamiltonian* system of $\overline{H}_3$ on the surface $\overline{H}_3=0$ are periodic. Considering complex coordinates of the corresponding period function, *Duistermaat* finds infinite branching. This excludes the existence of a third analytic integral.

In applications, very often assumptions arise which induce certain symmetries in the *Hamiltonian*. Such symmetries cause special bifurcations and other phenomena which are of practical interest. We discuss here some of the consequences of the assumption of discrete (mirror) symmetry in the position variable $q$.

**Discrete symmetry**

First we consider the case of discrete symmetry in $p_1,q_1$ or $p_3,q_3$ (or both). In the normal form we have $a_3=0$, since the *Hamiltonian* has to be invariant under M, defined by

$$M\phi_i=\phi_i+\pi \ , \ i=1,3.$$

Analysis of the critical points of the averaged equation shows that no periodic orbits in general position exist. There are still 7 periodic orbits, but the four in general position have moved into the $\tau_1=0$ and $\tau_3=0$ hyperplanes; see the action simplex in Figure 7.5.4-2.

Although this symmetry assumption reduces the number of terms in the normalized *Hamiltonian*, a third analytic integral does not exist in this case either. This can be deduced by using the analysis of *Duistermaat*(Dui84b) for this particular case.

We assume now discrete symmetry in $p_2,q_2$. This assumption turns out to have drastic consequences: the normal form to the third degree

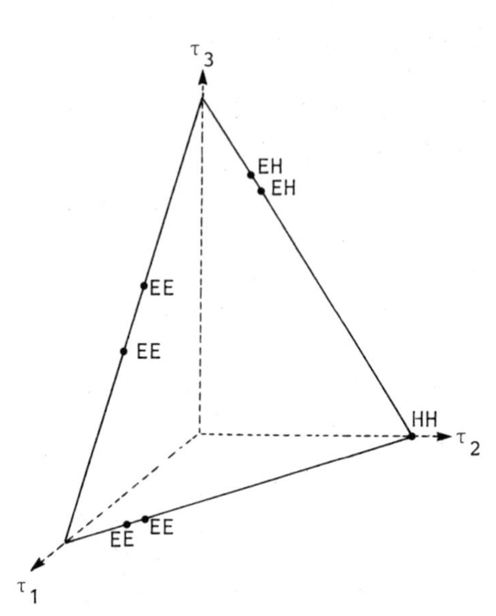


**Figure 7.5.4-2** *Action simplex for the discrete symmetric* 1:2:1*-resonance.*

vanishes, $\overline{H}_3=0$. So in this case, higher order averaging has to be carried out and the natural time-scale of the phase flow is at least of order $\frac{1}{\epsilon^2}$. The second order normal form contains one combination angle $\psi=2(\phi_1-\phi_3)$; this means that the resonance is not genuine, and that $\tau_2$ is a third integral of the system.

### 7.5.5. The 1:2:2 resonance

This case contains a surprise: *Martinet*, *Magnenat* and *Verhulst* (Mar81a) showed that the first order normalized system is integrable. This result was suggested by the consideration of numerically obtained stereoscopic projections of the flow in phase space. It is easy to generalize this result to the general *Hamiltonian* (Aa,83a). The normal from to first order is

$$\overline{H}=\tau_1+2\tau_2+2\tau_3+2\epsilon\tau_1[a_1(2\tau_2)^{1/2}\cos(2\phi_1-\phi_2-a_2)+a_3(2\tau_3)^{1/2}\cos(2\phi_1-\phi_3-a_4)],$$

where $a_i\in\mathbf{R}, i=1,\ldots,4$. Using the combination angles

$$2\psi_1=2\phi_1-\phi_2-a_2,$$

$$2\psi_2=2\phi_1-\phi_3-a_4,$$

we find the equations of motion

$$\dot{\tau}_1=4\epsilon\tau_1[a_1(2\tau_2)^{1/2}\sin(2\psi_1)+a_3(2\tau_3)^{1/2}\sin(2\psi_2)],$$

$$\dot{\tau}_2=-2\epsilon a_1\tau_1(2\tau_2)^{1/2}\sin(2\psi_1),$$

$$\dot{\tau}_3=-2\epsilon a_3\tau_1(2\tau_3)^{1/2}\sin(2\psi_2),$$

$$\dot{\psi}_1=\epsilon a_1(2\tau_2)^{-1/2}(4\tau_2-\tau_1)\cos(2\psi_1)+2\epsilon a_3(2\tau_3)^{1/2}\cos(2\psi_2),$$

$$\dot{\psi}_2 = 2\epsilon a_1(2\tau_2)^{\frac{1}{2}}\cos(2\psi_1) + \epsilon a_3(2\tau_3)^{-\frac{1}{2}}(4\tau_3 - \tau_1)\cos(2\psi_2).$$

Analyzing the critical points of the equations of motion we find in the energy plane $\tau_1 + 2\tau_2 + 2\tau_3 = constant$:

2 normal modes ($\tau_2$ and $\tau_3$ direction) which are unstable;

2 general position orbits which are stable;

1 global bifurcation set in the hyperplane $\tau_1 = 0$ (all solutions with $\tau_1 = 0$ are periodic in the first order normalized system).

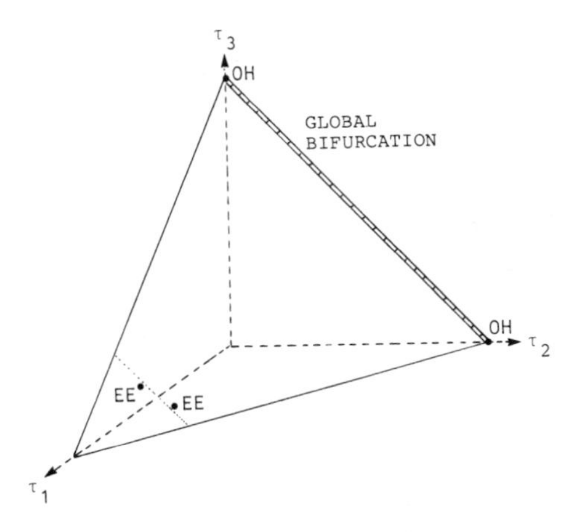


**Figure 7.5.5-1** *Action simplex for the* 1:2:2-*resonance normalized to* $H_3$. *The global bifurcation at* $\tau_1 = 0$ *corresponds with a continuous set of periodic solutions of the normalized Hamiltonian.*

Note that the phenomenon of a global bifurcation may not be stable under perturbation by higher order terms.

Apart from $H_2$ and $H_3$ we find a third integral, a quadratic one. The existence of this integral and the existence of the global bifurcation set are both tied in with the symmetry of the first order normal form. According to *Cushman* (Cus85a) the system splits after a suitable linear transformation into a 1:2-resonance and a one-dimensional subsystem.

*Van der Aa* and *Verhulst* (Aa,84a) considered two types of perturbation of the normal form to study the genericity of these phenomena. First, a simple deformation is obtained by detuning the resonance. Replace $H_2$ by

$$H_2 = \tau_1 + (2+\Delta_1)\tau_2 + (2+\Delta_2)\tau_3.$$

They find that in general no quadratic or cubic third integral exists in this case.

Secondly they consider how the global bifurcation and the integrability break up on adding higher order terms to the expansion of the normal form. In particular they consider the following asymmetric problem

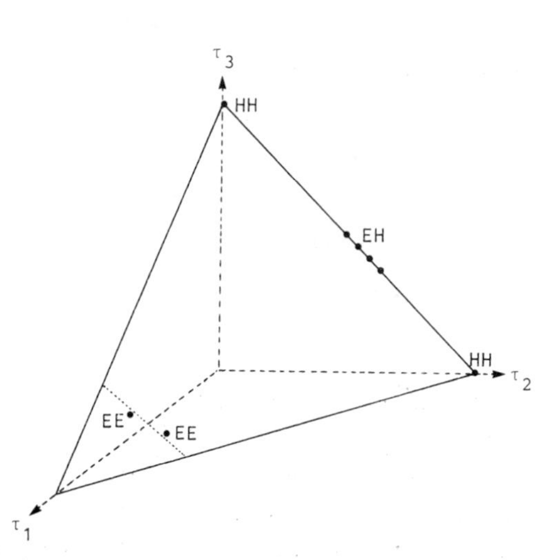


**Figure 7.5.5-2** *Action simplex of the* 1:2:2*-resonance normalized to* $H_4$*. The global bifurcation at* $\tau_1=0$ *has broken up into two normal modes and four periodic solutions with* $\tau_2\tau_3\neq 0$.

$$H_3=\tau_1+2\tau_2+2\tau_3+\epsilon(a_1q_1^2q_2+a_2q_1^2q_3+a_3q_1q_2q_3).$$

The parameter $a_3$ is the deformation parameter. From the point of view of applications this is a natural approach since it reflects approximate symmetry in a problem, which seems to be quite common. The global bifurcation is seen to break up into 6 periodic solutions (including the two normal modes). No third integral could be found in this case.

### 7.5.6. The 1:2:3 resonance

The general *Hamiltonian* was studied by *van der Aa* in (Aa,83a). Some aspects of the integrability of the 1:2:3 resonance were discussed by *Ford* in(For75a); *Kummer* (Kum75a) obtained periodic solutions using the normal form while comparing these with numerical results. More details can be found in appendix 8.5.

The first order normal form of the *Hamiltonian* is

$$H=\tau_1+2\tau_2+3\tau_3$$
$$+2\epsilon(2\tau_1\tau_2)^{1/2}[a_1\tau_3^{1/2}\cos(\phi_1+\phi_2-\phi_3-a_2)+a_3\tau_1^{1/2}\cos(2\phi_1-\phi_2-a_4)],$$

where, as usual, the $a_i, i=1,\ldots,4$ are constants. Introducing the combination angles

$$\psi_1=\phi_1+\phi_2-\phi_3-a_2,$$
$$\psi_2=2\phi_1-\phi_2-a_4$$

we find the equations of motion

$$\dot\tau_1=2\epsilon(2\tau_1\tau_2)^{1/2}[a_1\tau_3^{1/2}\sin(\psi_1)+2a_3\tau_1^{1/2}\sin(\psi_2)],$$
$$\dot\tau_2=2\epsilon(2\tau_1\tau_2)^{1/2}[a_1\tau_3^{1/2}\sin(\psi_1)-a_3\tau_1^{1/2}\sin(\psi_2)],$$

$$\dot{\tau}_3 = -2\epsilon a_1(2\tau_1\tau_2\tau_3)^{1/2}\sin(\psi_1),$$

$$\dot{\psi}_1 = 2\epsilon(2\tau_1\tau_2\tau_3)^{-1/2}[a_1(\tau_1\tau_3+\tau_2\tau_3-\tau_1\tau_2)\cos(\psi_1)+a_3(\tau_1\tau_3)^{1/2}(\tau_1+2\tau_2)\cos(\psi_2)],$$

$$\dot{\psi}_2 = \epsilon(\tau_1\tau_2)^{-1/2}[a_1(2\tau_3)^{1/2}(2\tau_2-\tau_1)\cos(\psi_1)+a_3(2\tau_1)^{1/2}(4\tau_2-\tau_1)\cos(\psi_2)].$$

Analyzing the critical points of the equation of motion we find 7 periodic solutions:

2 unstable normal modes ($\tau_2$ and $\tau_3$ direction);

1 stable solution in the $\tau_2=0$ hyperplane;

4 orbits in general position, two of which are stable and two unstable.

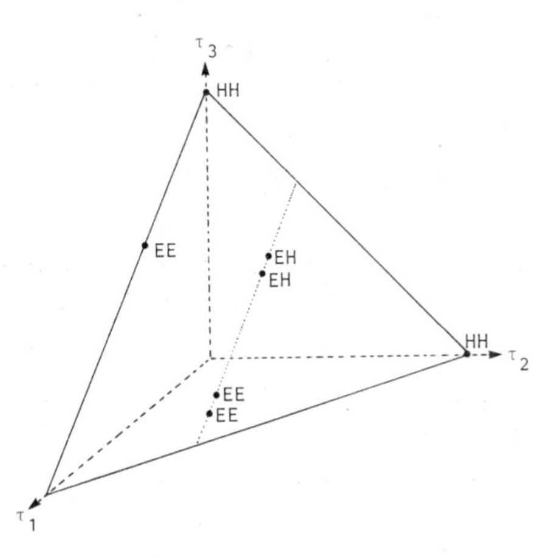


**Figure 7.5.6-1** *Action simplex for the* 1:2:3-*resonance.*

Apart from the usual two integrals no other integrals are known; *Van der Aa* has shown that no quadratic or cubic third integral exists.

**Discrete symmetry**

Discrete symmetry assumptions introduce drastic changes; we discuss the integrability aspects.

Mirror symmetry in $p_1,q_1$ or $p_3,q_3$ (or both) produces $a_1=0$. From the equations of motion we find that $\tau_3$ is constant, i.e. the system splits into two invariant subsystems: between the first and second degree of freedom we have a 1:2- resonance, in the third degree of freedom we have a nonlinear oscillator. So the system is clearly integrable with $\tau_3$ as the third integral. One can show that these results carry through for the second order normalized system.

Discrete symmetry in $p_2,q_2$ implies $a_1=a_3=0$, i.e. $\overline{H}_3=0$ (a similar strong degeneration of the normal form has been discussed in the section on the 1:2:1-resonance). Normalizing to second order produces a system which splits into a two-dimensional and a one-dimensional subsystem.

### 7.5.7. The 1:2:4 resonance

The normal form of this resonance (up till first order) is

$$\overline{H}=\tau_1+2\tau_2+4\tau_3$$
$$+2\epsilon[a_1\tau_1(2\tau_2)^{1/2}\cos(2\phi_1-\phi_2-a_2)+a_3\tau_2(2\tau_3)^{1/2}\cos(2\phi_2-\phi_3-a_4)],$$

where $a_1,\ldots,a_4\in\mathbf{R}$. This resonance has been studied by *Weinstein* (Wei78a) and, more detailed, by *van der Aa* (Aa,83a). Using combination angles

$$2\psi_1=2\phi_1-\phi_2-a_2,$$
$$2\psi_2=2\phi_2-\phi_3-a_4$$

the equations of motion become

$$\dot{\tau}_1=4\epsilon a_1\tau_1(2\tau_2)^{1/2}\sin(2\psi_1),$$
$$\dot{\tau}_2=-2\epsilon(2\tau_2)^{1/2}[a_1\tau_1\sin(2\psi_1)-2a_3(\tau_2\tau_3)^{1/2}\sin(2\psi_2)],$$
$$\dot{\tau}_3=-2\epsilon a_3\tau_2(2\tau_3)^{1/2}\sin(2\psi_2),$$
$$\dot{\psi}_1=\epsilon(2\tau_2)^{-1/2}[a_1(4\tau_2-\tau_1)\cos(2\psi_1)-2a_3(\tau_2\tau_3)^{1/2}\cos(2\psi_2)],$$
$$\dot{\psi}_2=\epsilon(2\tau_2\tau_3)^{-1/2}[2a_1\tau_1\tau_3^{1/2}\cos(2\psi_1)+a_3(4\tau_3-\tau_2)\tau_2^{1/2}\cos(2\psi_2)].$$

The analysis of periodic solutions differs slightly from the treatment of the preceding first order resonances as we have a bifurcation at the value $\Delta=16a_1^2-a_3^2=0$. From the analysis of the critical points we find:

1 unstable normal mode ($\tau_3$ direction)

If $\Delta<0$ we have 2 stable periodic solutions in the $\tau_1=0$ hyperplane, there are no periodic orbits in general position; at $\Delta=0$ two orbits branch off the $\tau_1=0$ solutions which for $\Delta>0$ become stable orbits in general position, while the $\tau_1=0$ solutions are unstable.

See figure 7.5.7-1

Apart from $H_2$ and $H_3$ no other independent integral of the normal system has been found, but it has been shown in (Aa,83a) that no quadratic or cubic integral exists.

Discrete symmetry in $p_1,q_1$ does not make the system integrable.

**Discrete symmetry**

Discrete symmetry in $p_2,q_2$ forces $a_1$ to be zero, and we consequently have a third integral $\tau_1$, producing the usual splitting into a one and a two degree of freedom system. These results carry over to second order normal forms.

Discrete symmetry in $p_3,q_3$ produces $a_3=0$ and the third integral $\tau_3$. Again we have the usual splitting, moreover the results carry over to second order.

Of course, the normal form degenerates if we assume discrete

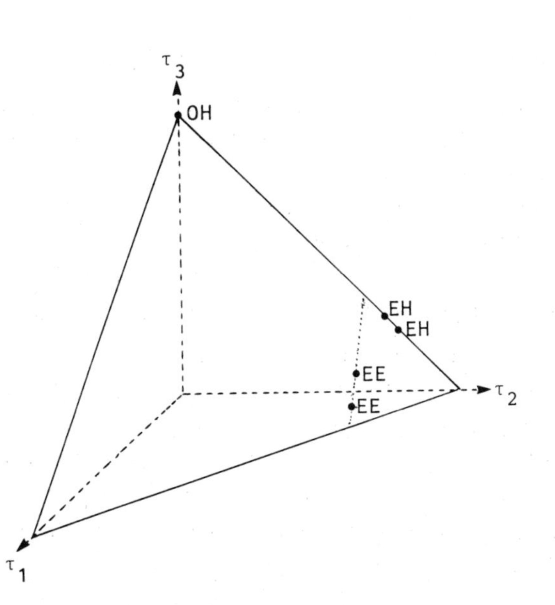


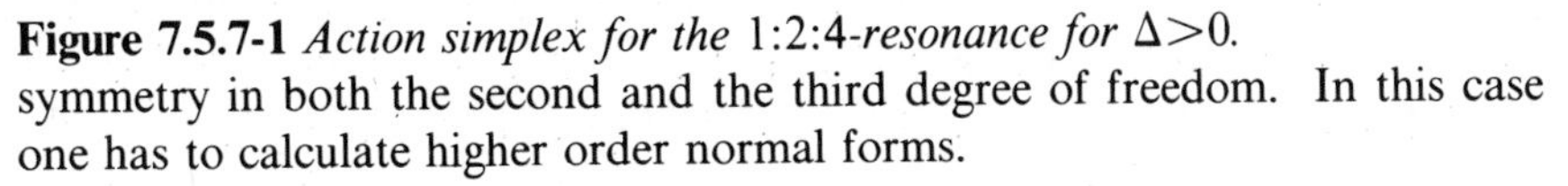

**Figure 7.5.7-1** *Action simplex for the* 1:2:4-*resonance for* $\Delta > 0$.

symmetry in both the second and the third degree of freedom. In this case one has to calculate higher order normal forms.

### 7.5.8. Non-genuine first order resonance

Applications in the theory of vibrating systems sometimes produces first order resonances. Suppose one considers the interaction of three modes, characterized by the frequencies 1, 2 and $\omega$. We take $\omega \neq 1,2,3,4$ to exclude the genuine first order resonances treated earlier. Normalization to $H_3$ produces $(a_1, a_2 \in \mathbf{R})$

$$\overline{H} = \tau_1 + 2\tau_2 + \omega\tau_3 + 4\epsilon a_1 \tau_1 \tau_2^{1/2} \cos(2\phi_1 - \phi_2 - a_2)$$

which clearly exhibits the 1:2-resonance between the first two modes while $\tau_3$ is constant. There are three independent integrals of the normalized system. From § 7.4.3 we have in the 1:2-resonance two stable periodic orbits in general position and one hyperbolic normal mode. Adding the third mode, we have three families of periodic solutions for each value of the energy; see Figure 7.4.3-1

To study the interaction between all three modes, we have to normalize to higher order. To study these phenomena *van der Aa* considered in (Aa,83a) the case $\omega = 5$; an annihilation-vector is then $(2,1,-1)$ so that we expect nontrivial results upon normalization to $H_4$. Introducing the real constants $b_1, \ldots, b_8$ and the combination angles

$$\psi_1 = 2\phi_1 - \phi_2 - a_2,$$

$$\psi_2 = \phi_1 + 2\phi_2 - \phi_3 - b_8$$

we have

$$\overline{H} = \tau_1 + 2\tau_2 + 5\tau_3 + 4\epsilon a_1 \tau_1 \tau_2^{1/2} \cos(\psi_1) + 4\epsilon^2 [b_1 \tau_1^2 + b_2 \tau_1 \tau_2$$

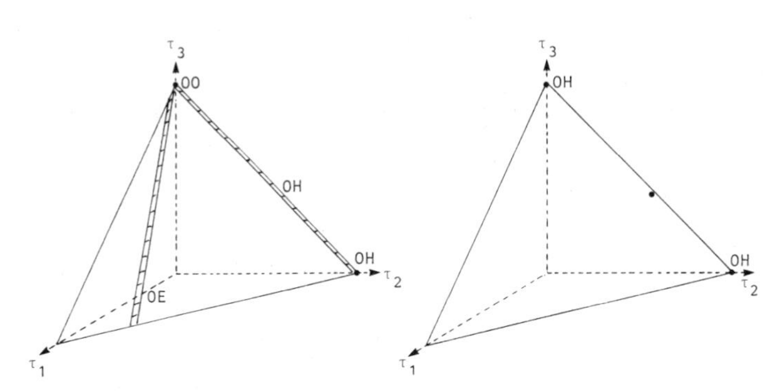


**Figure 7.5.8-1** *Action simplices for the* 1:2:5*-resonance normalized to* $H_3$ *and to* $H_4$. *The normalization to* $H_4$ *produces a break-up of the two global bifurcations.*

$$+b_3\tau_1\tau_3+b_4\tau_2^2+b_5\tau_2\tau_3+b_6\tau_3^2+b_7\tau_2(\tau_1\tau_3)^{1/2}\cos(\psi_3)].$$

The analysis of the equations of motion gives surprising results. The two families of stable orbits in general position vanish on adding $\overline{H}_4$. The normal mode family $\tau_1=0$, $\tau_3=$*constant* breaks up as follows: in the hyperplane $\tau_1=0$ we have two normal modes $\tau_2=0$ resp. $\tau_3=0$ and a family of periodic orbits with $\tau_1\tau_3>0$; the normal modes are hyperbolic, the stability of the family of periodic solutions depends on the parameters. The results are illustrated in Figure 7.5.8-1.

### 7.5.9. Summary of integrability of normalized systems

We summarize the results from the preceding sections on three degrees freedom systems with respect to integrability after normalization in Table III. If three independent integrals of the normalized system can be found, the normalized system is integrable; the original system is in this case called formally integrable. The integrability depends in principle on how far the normalization is carried out. The formal integrals have a precise asymptotic meaning, see § 7.4.3. We have the following abbreviations: "no cubic integral" for no quadratic or cubic third integral; "discr. symm. $\underline{q_i}$" for discrete symmetry in the $p_i,q_i$-degree of freedom; "2 subsystems at $H_k$" for the case that the normalized system decouples into a one and a two degrees of freedom subsystem upon normalizing to $H_k$. In the second and third column one finds the number of known integrals including $\overline{H}_3$ and $\overline{H}_4$, respectively.

The remarks which have been added to the table reflect some of the results known on the non-existence of a third integral. Note that the results presented here are for the general *Hamiltonian* and that additional assumptions may change the results.

| Resonance | | $H_3$ | $H_4$ | Remarks |
|---|---|---|---|---|
| *1:2:1* | *general* | 2 | 2 | *no analytic third integral* |
| | *discr.symm.* $q_1$ | 2 | 2 | *no analytic third integral* |
| | *discr.symm.* $q_2$ | 3 | 3 | $\overline{H}_3=0$; *2 subsystems at* $\overline{H}_4$ |
| | *discr.symm.* $q_3$ | 2 | 2 | *no analytic third integral* |
| *1:2:2* | *general* | 3 | 2 | *no cubic integral at* $\overline{H}_4$ |
| | *discr.symm.* $q_2$ *and* $q_3$ | 3 | 3 | $\overline{H}_3=0$; *2 subsystems at* $\overline{H}_4$ |
| *1:2:3* | *general* | 2 | 2 | *no cubic integral* |
| | *discr.symm.* $q_1$ | 3 | 3 | *2 subsystems at* $\overline{H}_3$ *and* $\overline{H}_4$ |
| | *discr.symm.* $q_2$ | 3 | 2 | $\overline{H}_3=0$ |
| | *discr.symm.* $q_3$ | 3 | 3 | *2 subsystems at* $\overline{H}_3$ *and* $\overline{H}_4$ |
| *1:2:4* | *general* | 2 | 2 | *no cubic integral* |
| | *discr.symm.* $q_1$ | 2 | 2 | *no cubic integral* |
| | *discr.symm.* $q_2$ *or* $q_3$ | 3 | 3 | *2 subsystems at* $\overline{H}_3$ *and* $\overline{H}_4$ |

**Table III**

**References**


Aa,79a. van der Aa,E. and Sanders,J.A., "On the 1:2:1- resonance, its periodic orbits and integrals," in *Asymptotic Analysis, from Theory to Application*, ed. F.Verhulst, Lecture Notes in Mathematics #711, Springer-Verlag, Berlin (1979).

Aa,83a. van der Aa,E., "First order resonances in three-degrees-of-freedom systems," *Celestial Mechanics* **31**, pp. 163-191 (1983).

Aa,84a. van der Aa,E. and Verhulst,F., "Asymptotic integrability and periodic solutions of a *Hamiltonian* system in 1:2:2-resonance," *SIAM J. Math. Anal.* **15**, pp. 890-911 (1984).

Abr78a. Abraham,R. and Marsden,J.E., *Foundations of Mechanics (2nd edition),* The Benjamin/Cummings Publ. Co., Reading, Mass. (1978).

Abr83a. Abraham,R.H. and Shaw,C.D., *Dynamics-The Geometry of Behavior I,II,* Aerial Press, Inc., Santa Cruz, California (1983).

Arn64a. Arnold,V.I., "Instability of dynamical systems with several degrees of freedom," *Dokl. Akad. Nauk. SSSR* **156**, pp. 581-585 (1964).

Arn65a. Arnold,V.I., "Conditions for the applicability, and estimate of the error, of an averaging method for systems which pass through states of resonance during the course of their evolution," *Soviet Math.* **6**, pp. 331-334 (1965).

Arn78a. Arnold,V.I., *Mathematical Methods of Classical Mechanics,* Springer-Verlag, New - York (1978).

Arn83a. Arnold,V.I., *Geometrical Methods in the Theory of Ordinary Differential Equations,* Springer-Verlag, New York (1983).

Bal75a. Balachandra,M. and Sethna,P.R., "A generalization of the method of averaging for systems with two time-scales," *Archive Rat. Mech.*

*Anal.* **58**, pp. 261-283 (1975).

Bal82a. Balbi,J.H., "Averaging and reduction," *Int. J. Non-linear Mechanics* **17**(5/6), pp. 343-353 (1982).

Ban67a. Banfi,C., "Sull'approssimazione di processi non stazionari in meccanica non lineare," *Bolletino dell Unione Matematica Italiana* **22**, pp. 442-450 (1967).

Ban69a. Banfi,C. and Graffi,D., "Sur les méthodes approchées de la mécanique non linéaire," *Actes du Coll. Equ. Diff. Non Lin.*, Mons, pp. 33-41 (1969).

Ber78a. Berry,M.V., "Regular and irregular motion," pp. 16-120 in *Topics in Nonlinear Dynamics*, ed. Jorna,S., Am. Inst. Phys. Conf. Proc. (1978).

Bes69a. Besjes,J.G., "On the asymptotic methods for non-linear differential equations," *Journal de Mécanique* **8**, pp. 357-373 (1969).

Bog61a. Bogoliubov,N.N. and Mitropolskii,Yu.A., *Asymptotic methods in the theory of nonlinear oscillations,* Gordon and Breach, New York (1961).

Bog76a. Bogoliubov,N.N., Mitropolskii,Yu.A., and Samoilenko,A.M., *Methods of Accelerated Convergence in Nonlinear Mechanics,* Hindustan Publ. Co. and Springer Verlag, Delhi and Berlin (1976).

Boh32a. Bohr,H., *Fastperiodische Funktionen,* Springer Verlag, Berlin (1932).

Bur74a. van der Burgh,A.H.P., *Studies in the Asymptotic Theory of Nonlinear Resonance,* Technical Univ., Delft (1974).

Byr71a. Byrd,P.F. and Friedman,M.B., *Handbook of Elliptic Integrals for Engineers and Scientists,* Springer Verlag (1971).

Cap73a. Cap,F.F., "Averaging method for the solution of non-linear differential equations with periodic non-harmonic solutions," *International Journal Non-linear Mechanics* **9**, pp. 441-450 (1973).

Cod55a. Coddington,A. and Levinson,N., *Theory of Ordinary Differential Equations,* McGraw-Hill, New-York (1955).

Cus82a. Cushman,R., "Reduction of the 1:1 nonsemisimple resonance," *Hadronic Journal* **5**, pp. 2109-2124 (1982).

Cus85a. Cushman,R.H., "1:2:2 resonance ," manuscript (1985).

Dui84a. Duistermaat,J.J., "Bifurcations of periodic solutions near equilibrium points of *Hamiltonian* systems," in *Bifurcation Theory and Applications*, ed. L.Salvadori, Lecture Notes in Mathematics #1057, Springer-Verlag (1984).

Dui84b. Duistermaat,J.J., "Non-integrability of the 1:1:2-resonance," *Ergodic Theory and Dynamical Systems* **4**, pp. 553-568 (1984).

Dui84c. Duistermaat,J.J., "Erratum to: Non-integrability of the 1:1:2-resonance," Preprint (1984).

Eck75a. Eckhaus,W., "New approach to the asymptotic theory of nonlinear oscillations and wave-propagation," *J. Math. An. Appl.* **49**, pp. 575-611 (1975).

Eck79a. Eckhaus,W., *Asymptotic Analysis of Singular Perturbations,* North-Holland Publ. Co., Amsterdam (1979).

Eul54a. Euler,L., "De seribus divergentibus," *Novi commentarii ac. sci. Petropolitanae* **5**, pp. 205-237 (1754).

Eul24a. Euler,L., pp. 585-617 in *Opera Omnia, ser. I,14* (1924).

Eva76a. Evan-Iwanowski,R.M., *Resonance Oscillations in Mechanical Systems,* Elsevier Publ. Co., Amsterdam (1976).

Fin74a. Fink,A.M., *Almost Periodic Differential Equations,* Lecture Notes in Mathematics 377, Springer Verlag, Berlin (1974).

For75a. Ford,J., "Ergodicity for nearly linear oscillator systems," in *Fundamental Problems in Statistical Mechanics III*, ed. Cohen,E.G.D. (1975).

Fra69a. Fraenkel,L.E., "On the method of matched asymptotic expansions I,II,III," *Proc. Phil. Soc.* **65**, pp. 209-284 (1969).

Gol71a. Goloskokow,E.G. and Filippow,A.P., *Instationäre Schwingungen Mechanischer Systeme,* Akademie Verlag, Berlin (1971).

Gre84a. Greenspan,B. and Holmes,Ph., "Repeated resonance and homoclinic bifurcation in a periodically forced family of oscillators," *SIAM J. Math. Analysis* **15**, pp. 69-97 (1984).

Guc83a. Guckenheimer,J. and Holmes,P., *Nonlinear Oscillations, Dynamical Systems, and Bifurcations of Vector Fields,* Springer Verlag, Applied Mathematical Sciences, New York (1983).

Had63a. Hadjidemetriou,J.D., "Two-body problem with variable mass: a new approach," *Icarus* **2**, p. 440 (1963).

Hal69a. Hale,J.K., *Ordinary Differential Equations,* Wiley-Interscience, New York (1969).

Hec81a. Hecke, *Lectures on the Theory of Algebraic Numbers,* Springer-Verlag, New York (1981).

Hel80a. Helleman,R.H.G., "Self-generated chaotic behaviour in nonlinear mechanics," pp. 165-233 in *Fundamental Problems in Statistical Mechanics*, ed. Cohen,E.G.D., North Holland Publ., Amsterdam and New York (1980).

Jea28a. Jeans,J.H., *Astronomy and Cosmogony,* At The University Press, Cambridge (1928).

Kev74a. Kevorkian,J., "On a model for reentry roll resonance," *SIAM J. Appl. Math.* **35**, pp. 638-669 (1974).

Kir78a. Kirchgraber,U. and Stiefel,E., *Methoden der analytischen Störungsrechnung und ihre Anwendungen,* B.G.Teubner, Stuttgart (1978).

Kum75a. Kummer,M., "An interaction of three resonant modes in a nonlinear lattice," *Journal of Mathematical Analysis and Applications* **52**, pp. 64-104 (1975).

Lic83a. Lichtenberg,A.J. and Lieberman,M.A., *Regular and Stochastic Motion,* Springer Verlag, Applied Mathematical Sciences, New York (1983).

Lya47a. Lyapunov,A.M., "Problème general de la stabilité du mouvement," *Ann. of Math. Studies*, Princeton,N.J. **17**, Princeton Univ. Press (1947).

Mar81a. Martinet,L., Magnenat,P., and Verhulst,F., "On the number of isolating integrals in resonant systems with 3 degrees of freedom," *Celestial Mechanics* **29**, pp. 93-99 (1981).

Mee82a. van der Meer,J.-C., "Nonsemisimple 1:1 resonance at an equilibrium," *Celestial Mechanics* **27**, pp. 131-149 (1982).

Mey84a. Meyer,K.R., "Normal forms for the general equilibrium," *Funkcialaj ekvacioj* **27**(2), pp. 261-271 (1984).

Mit65a. Mitropolsky,Ya.A., *Problems of the Asymptotic Theory of Nonstationary Vibrations,* Israel Progr. Sc. Transl., Jerusalem (1965).

Mit73a. Mitropolsky,Ya.A., *Certains aspects des progrès de la méthode de centrage,* CIME, Edizione Cremonese, Roma (1973).

Mos71a. Moser,J. and Siegel,C.L., *Lectures on Celestial Mechanics,* Springer-Verlag (1971).

Nek77a. Nekhoroshev,N.N., "An exponential estimate of the time of stability of nearly-integrable *Hamiltonian* systems," *Usp. Mat. Nauk* **32**(6), pp. 5-66 (1977).

Per69a. Perko,L.M., "Higher order averaging and related methods for perturbed periodic and quasi-periodic systems," *SIAM J. Appl. Math.* **17**, pp. 698-724 (1969).

Poi93a. Poincaré,H., *Les Méthodes Nouvelles de la Mécanique Céleste II,* Gauthiers-Villars, Paris (1893).

Rob81a. Robinson,C., "Stability of periodic solutions from asymptotic expansions," pp. 173-185 in *Classical Mechanics and Dynamical Systems,* ed. Devaney,R.L. and Nitecki,Z.H., Marcel Dekker, Inc., New York (1981).

Rob83a. Robinson,C., "Sustained resonance for a nonlinear system with slowly varying coefficients," *SIAM J. Math. Analysis* **14**(5), pp. 847-860 (1983).

Ros66a. Roseau,M., *Vibrations nonlinéaires et théorie de la stabilité,* Springer-Verlag (1966).

Ros70a. Roseau,M., "Solutions périodiques ou presque périodiques des systèmes differentiels de la mécanique non-linéaire," in *I.C.S.M. Course 44, Undine*, Springer Verlag, Wien-New York (1970).

San75a. Sanchez-Palencia,E., "Méthode de centrage et comportement des trajectoires dans l'espace des phases," *Compt. Rend. Acad. Sci., Ser. A*,

Paris **280**, pp. 105-107 (1975).

San76a. Sanchez-Palencia,E., "Methode de centrage - estimation de l'erreur et comportement des trajectoires dans l'espace des phases," *Int. J. Non-Linear Mechanics* **11**(176), pp. 251-263 (1976).

San78a. Sanders,J.A., "Are higher order resonances really interesting?," *Celestial Mechanics* **16**, pp. 421-440 (1978).

San78b. Sanders,J.A., "On the Fermi-Pasta-Ulam chain," Preprint #74,RUU (1978).

San79a. Sanders,J.A., "On the passage through resonance," *SIAM J. Math. An.* **10**, pp. 1220-1243 (1979).

San80a. Sanders,J.A., "Asymptotic approximations and extension of time-scales," *SIAM J. Math. An.* **11**, pp. 758-770 (1980).

San82a. Sanders,J.A., "Melnikov's method and averaging," *Celestial Mechanics* **28**, pp. 171-181 (1982).

Slu70a. van der Sluis,A., "Domains of uncertainty for perturbed operator equations," *Computing* **5**, pp. 312-323 (1970).

Ver75a. Verhulst,F., "Asymptotic expansions in the perturbed two-body problem with application to systems with variable mass," *Celestial Mechanics* **11**, pp. 95-129 (1975).

Ver76a. Verhulst,F., "On the theory of averaging," pp. 119-140 in *Long Time Predictions in Dynamics*, ed. Szebehely,V. and Tapley,B.D., Reidel, Dordrecht (1976).

Ver79a. Verhulst,F., "Discrete symmetric dynamical systems at the main resonances with applications to axi-symmetric galaxies," *Philosophical Transactions of the Royal Society of London* **290**, pp. 435-465 (1979).

Ver83a. Verhulst,F., "Asymptotic analysis of *Hamiltonian* systems," *Lecture Notes in Mathematics* **985**, pp. 137-183, Springer-Verlag (1983).

Vol63a. Volosov,V.M., "Averaging in systems of ordinary differential equations," *Russ. Math. Surveys* **17**, pp. 1-126 (1963).

Wei73a. Weinstein,A., "Normal modes for nonlinear *Hamiltonian* systems," *Inv. Math.* **20**, pp. 47-57 (1973).

Wei78a. Weinstein,A., "Simple periodic orbits," in *AIP Conf. Proc.* (1978).

# 8. Appendices

## 8.1. The History of the Theory of Averaging

### 8.1.1. Early calculations and ideas

Perturbation methods for differential equations became important when scientists in the $18^{th}$ century were trying to relate *Newton*'s theory of gravitation to the observations of the motion of planets and satellites. Right from the beginning it became clear that a dynamical theory of the solar system based on a superposition of only two-body motions, one body being always the sun and the other body being formed by the respective planets, produces a reasonable but not very accurate fit to the observations. To explain the deviations one considered effects as the influence of satellites like the moon in the case of the earth, the interaction of large planets like *Jupiter* and *Saturn*, the resistance of the ether and other effects. These considerations led to the formulation of perturbed two-body motion and, as exact solutions were clearly not available, the development of perturbation theory.

The first attempts took place in the first half of the $18^{th}$ century and involve a numerical calculation of the increments of position and velocity variables from the differentials during successive small intervals of time. The actual calculations involve various ingenious expansions of the perturbation terms to make the process tractable in practice. It soon became clear that this process leads to the construction of astronomical tables but not necessarily to general insight into the dynamics of the problem. Moreover the tables were not very accurate as to obtain high accuracy one has to take very small intervals of time. An extensive study of early perturbation theory and the construction of astronomical tables has been presented by *Wilson* (Wil80a) and the reader is referred to this work for details and references.

New ideas emerged in the second half of the 18$^{\text{th}}$ century by the work of *Clairaut*, *Lagrange* and *Laplace*. It is difficult to settle priority claims as the scientific gentlemen of that time did not bother very much with the acknowledgment of ideas or references. It is clear however that *Clairaut* had some elegant ideas about particular problems at an early stage and that *Lagrange* was able to extend and generalize this considerably, while presenting the theory in a clear and to the general public understandable way.

*Clairaut* (Cla54a) wrote the solution of the (unperturbed) two-body problem in the form

$$\frac{p}{r} = 1 - c\cos(v),$$

where $r$ is the distance of the two bodies, $v$ the longitude measured from aphelion, $p$ is a parameter, $c$ the eccentricity of the conic section. Admitting a perturbation $\Omega$, *Clairaut* derives the integral equation by a variation of constants procedure; he finds

$$\frac{p}{r} = 1 - c\cos(v) + \sin(v)\int \Omega\cos(u)du - \cos(v)\int \Omega\sin(u)du.$$

The perturbation $\Omega$ depends on $r$, $v$ and maybe other quantities; in the expression for $\Omega$ we replace $r$ by the solution of the unperturbed problem and we assume that we may expand in cosines of multiples of $v$

$$\Omega = A\cos(av) + B\cos(bv) + \cdots .$$

The perturbation part of *Clairaut*'s integral equation contains upon integration terms like

$$-\frac{A}{a^2-1}\cos(av) - \frac{B}{b^2-1}\cos(bv), \qquad a,b \neq 1.$$

If the series for the perturbation term $\Omega$ contains a term of the form $\cos(v)$, integration yields terms like $v\sin(v)$ which represent secular (unbounded) behavior of the orbit. In this case *Clairaut* adjusts the expansion to eliminate this effect.

Although this process of calculating perturbation effects is not what we call averaging now, it has some of its elements. First there is the technique of integrating while keeping slowly varying quantities like $c$ fixed; secondly there is a procedure to avoid secular terms which is related to the modern approach (see appendix 8.2).

This technique of obtaining approximate solutions is developed and is used extensively by *Lagrange* and *Laplace*. The treatment in *Laplace*'s 'Traité de Mécanique Céleste' is however very technical and the underlying ideas are not presented to the reader in a comprehensive way. One can find the ingredients of the method of averaging and also higher order perturbation procedures in *Laplace*'s study of the *Sun-Jupiter-Saturn* configuration;

see for instance (Lap25a), book 2, chapter 5-8 and book 6. We shall turn now to the expositions of *Lagrange* who describes the perturbation method employed in his work in a transparent way.

Instead of referring to various papers by *Lagrange* we cite from the Mécanique Analytique, published in 1788. After discussing the formulation of motion in dynamics *Lagrange* argues that to analyze the influence of perturbations one has to use a method which we now call 'variation of parameters'. The start of the 2nd part, 5th section, art. 1 reads in translation:

> "1. All approximations suppose (that we know) the exact solution of the proposed equation in the case that one has neglected some elements or quantities which one considers very small. This solution forms the first order approximation and one improves this by taking successively into account the neglected quantities.
>
> In the problems of mechanics which we can only solve by approximation one usually finds the first solution by taking into account only the main forces which act on the body; to extend this solution to other forces which one can call perturbations, the simplest course is to conserve the form of the first solution while making variable the arbitrary constants which it contains; the reason for this is that if the quantities which we have neglected and which we want to take into account are very small, the new variables will be almost constant and we can apply to them the usual methods of approximation. So we have reduced the problem to finding the equations between these variables."

*Lagrange* then continues to derive the equations for the new variables, which we now call the perturbation equations in the standard form. In art. 16 of the 2nd part, 5th section the decomposition is discussed of the perturbing forces in periodic functions which leads to averaging. In art. 20-24 of the same section a perturbation formulation is given which describes the variation of quantities as the energy. To illustrate the relation with averaging we give *Lagrange*'s discussion of secular perturbations in planetary systems. *Lagrange* introduces a perturbation term $\Omega$ in the discussion of the 2nd part, 7th section, art. 76. This reads in translation:

> "To determine the secular variations one has only to substitute for $\Omega$ the nonperiodic part of this function, i.e. the first term of the expansion of $\Omega$ in the sine and cosine series which depend on the motion of the perturbed planet and the perturbing planets. $\Omega$ is only a function of the elliptical coordinates of these planets and provided that the eccentricities and the inclinations are of no importance, we can always reduce these coordinates to a sine and cosine series in angles which are related to anomalies and average longitudes; so we can also expand the function $\Omega$ in a series of the same type and the first term which contains no sine or cosine will be the only one which can produce secular equations."
>
> Comparing the method of *Lagrange* with our introduction of the

averaging method in § 2.4-6, we note that *Lagrange* starts by transforming the problem to the standard form

$$\dot{x} = \epsilon f(t,x) + \epsilon^2 \cdots , \; x(0) = x_o$$

by 'variations des constantes.' Then the function $f$ is expanded in what we now call a *Fourier* series with respect to $t$, involving coefficients depending on $x$ only

$$\dot{x} = \epsilon f^o(x) + \epsilon \sum_{n=1}^{\infty} [a_n(x)\cos(nt) + b_n(x)\sin(nt)] + \epsilon^2 \cdots .$$

Keeping the first, time-independent term yields the secular equation

$$\dot{y} = \epsilon f^o(y) \, , \; y(0) = x_o .$$

This equation produces the secular changes of the solutions according to *Lagrange* 2nd part, section 45, § 3, art. 19. It is precisely the equation obtained by first order averaging as described in § 2.6. At the same time no unique meaning is attributed to what we call a first correction to the unperturbed problem. If $f^o = 0$ it sometimes means replacing in the equation $x$ by $x_o$ so that we have a first order correction like

$$x(t) = x_o + \int_0^t f(\tau, x) d\tau .$$

Sometimes the first order correction involves more complicated expressions. This confusion of terminology will last until in the 20$^{th}$ century definitions and proofs have been formulated.

## 8.1.2. Formal perturbation theory and averaging

Perturbation theory as developed by *Clairaut*, *Laplace* and *Lagrange* has been used from 1800 onwards as a collection of formal techniques. The theory can be traced in many 19$^{th}$ and 20$^{th}$ century books on celestial mechanics and dynamics; we shall discuss some of its aspects in the work of *Jacobi*, *Poincaré* and *van der Pol*. See also the book by *Born*(Bor27a).

### 8.1.2.1. Jacobi

The lectures of *Jacobi* (Jac69a) on dynamics show a remarkable development of the theoretical foundations of mechanics: the discussion of *Hamilton* equations of motion, the partial differential equations called after *Hamilton – Jacobi* and many other aspects. In Jacobi's 36$^{th}$ lecture on dynamics perturbation theory is discussed.

The main effort of this lecture is directed towards the use of *Lagrange*'s 'variation des constantes' in a canonical way. After presenting the unperturbed problem by *Hamilton*'s equations of motion, *Jacobi* assumes that the perturbed problem is characterized by a *Hamiltonian* function. If certain transformations are introduced, the perturbation equations in the standard form are shown to have again the same *Hamiltonian* structure. This formulation of what we now call canonical perturbation theory has many advantages and it has become the standard formulation in perturbation theory of *Hamiltonian* mechanics.

Note however that this treatment concerns only the way in which the standard perturbation form

$$\dot{x} = \epsilon f(t,x) + \epsilon^2 \cdots$$

is derived. It represents an extension of the first part of the perturbation theory of *Lagrange*. The second part, i.e. how to treat these perturbation equations, is discussed by *Jacobi* in a few lines in which the achievements of *Lagrange* are more or less ignored. About the introduction of the standard form involving the perturbation $\Omega$ *Jacobi* states (in translation):

> "This system of differential equations has the advantage that the first correction of the elements is obtained by simple quadrature. This is obtained on considering the elements as constant in $\Omega$ while giving them the values which they had in the unperturbed problem. Then $\Omega$ becomes simply a function of time $t$ and the corrected elements are obtained by simple quadrature. The determination of higher corrections is a difficult problem which we do not go into here."

*Jacobi* does not discuss why *Lagrange*'s secular equation is omitted in this *Hamiltonian* framework; in fact, his procedure is incorrect as we *do* need the secular equation for a correct description.

### 8.1.2.2. Poincaré

We shall base ourselves in this discussion on the two series of books written by *Poincaré* on celestial mechanics: 'Les méthodes nouvelles de la Mécanique Céleste' and the 'Leçons de Mécanique Céleste.' The first one, which we shall indicate by 'Méthodes,' is concerned with the mathematical foundations of celestial mechanics and dynamical systems; the second one, which we shall indicate by 'Leçons,' aims at the practical use of mathematical methods in celestial mechanics.

The 'Méthodes' is still a rich source of ideas and methods in mathematical analysis; we only consider here the relation with perturbation theory. In volume I of (Poi92a), chapter 3, *Poincaré* considers the determination of periodic solutions by series expansion with respect to a small parameter. Consider for instance the equation

$$\ddot{x} + x = \epsilon f(x,\dot{x})$$

and suppose that an isolated periodic solution exists for $0<\epsilon<<1$; if $\epsilon=0$ all solutions are periodic. Note that this example has some similarity with the case of perturbed *Kepler* motion. Under certain conditions *Poincaré* proves that we can describe the periodic solution by a *convergent* series in entire powers of $\epsilon$, where the coefficients are bounded functions of time. In volume II of the 'Méthodes,' *Poincaré* demonstrates the application of the method and, if the conditions have not been satisfied, its failures to produce convergent series. In the actual calculations *Poincaré* employs *Lagrange*'s and *Jacobi*'s perturbation formulation supplemented by a secularity condition which is justified for periodic solutions. The conditions which we do not discuss here, are connected with the possibility of continuation or branching of solutions.

It is interesting to note that in the 'Méthodes,' *Poincaré* has also justified the use of divergent series by the introduction of the concept of asymptotic series. It is this concept which nowadays enables us to give a precise meaning to series expansion by averaging methods.

The 'Leçons' is concerned with the actual application of the 'Méthodes' in celestial mechanics. The first volume deals with the theory of planetary perturbations and contains a very complete discussion of *Lagrange*'s secular perturbation theory (the theory of averaging); moreover the theory is added to by the study of many details and special cases. The approximations remain formal except in the case of periodic solutions.

### 8.1.2.3. Van der Pol

In the theory of nonlinear oscillations the method of *van der Pol* is concerned with obtaining approximate solutions for equations of the type

$$\ddot{x}+x=\epsilon f(x,\dot{x}).$$

In particular for the *van der Pol* equation we have

$$f(x,\dot{x})=(1-x^2)\dot{x}$$

which arises in studying triode oscillations (Pol26a). *Van der Pol* introduces the transformation $(x,\dot{x})\mapsto(a,\phi)$ by

$$x=a\sin(t+\phi),$$

$$\dot{x}=-a\cos(t+\phi).$$

The equation for $a$ can be written as

$$\frac{da^2}{dt}=\epsilon a^2(1-\tfrac{1}{4}a^2)+\cdots,$$

where the dots stand for higher order harmonics. Omitting the terms represented by the dots, as they have zero average, *van der Pol* obtains an equation which can be integrated to produce an approximation of the amplitude $a$.

Note that the transformation $x=a\sin(t+\phi)$ is an example of *Lagrange*'s 'variation des constantes.' The equation for the approximation of $a$ is the secular equation of *Lagrange* for the amplitude. Altogether *van der Pol*'s method is an interesting special example of the perturbation method described by *Lagrange* in (Lag88a).

One might wonder whether *van der Pol* realized that the technique which he employed is an example of classical perturbation techniques. The answer is very probably affirmative. *Van der Pol* graduated in 1916 at the University of Utrecht with main subjects physics, he defended his doctorate thesis in

1920 at the same university. In that period and for many years thereafter the study of mathematics and physics at the Dutch universities involved celestial mechanics which often contained some perturbation theory. A more explicit answer can be found in (Pol20a) on the amplitude of triode vibrations; on page 704 *van der Pol* states that the equation under consideration "is closely related to some problems which arise in the analytical treatment of the perturbations of planets by other planets." This seems to establish the relation of *van der Pol*'s analysis for triodes with celestial mechanics.

## 8.1.3. Proofs of asymptotic validity

The first proof of the asymptotic validity of the averaging method was given by *Fatou* (Fat28a). Assuming periodicity with respect to $t$ and continuous differentiability of the vectorfield, *Fatou* uses the *Picard-Lindelöf* iteration procedure to obtain $O(\epsilon)$ estimates on the time-scale $\frac{1}{\epsilon}$. The proof is based essentially on the iteration (contraction) results developed at the end of the $19^{th}$ century.

In the Soviet-Union similar results were obtained by *Mandelstam* and *Papalexi* (Man34a). An important step forward is the development and proof of the averaging method in the case of almost-periodic vectorfields by *Krylov* and *Bogoliubov* in (Kry37a). This is followed by *Bogoliubov*'s averaging results in the general case where for the equation

$$\dot{x} = \epsilon f(t,x).$$

*Bogoliubov* requires that the general average exists:

$$\lim_{T\to\infty} \frac{1}{T} \int_0^T f(t,x)dt.$$

An important part has been played by the monograph on nonlinear oscillations by *Bogoliubov* and *Mitropolsky* (Bog61a). The book has been very influential because of its presentation of both many examples and an elaborate discussion of the theory. An account of *Mitropolsky*'s theory for systems with coefficients slowly varying with time can also be found in this book.

The theory of averaging has been developed after this for many branches of nonlinear analysis. A transparent proof using *Gronwall*'s inequality for the case of periodic differential equations has been provided by *Roseau* (Ros66a). Some notes on the literature of new developments in the theory of differential equations have been given in § 3.1 of the present monograph.

### References

Bog61a. Bogoliubov,N.N. and Mitropolskii,Yu.A., *Asymptotic methods in the theory of nonlinear oscillations,* Gordon and Breach, New York (1961).

Bor27a. Born,M., *The mechanics of the atom,* G.Bell and Sons, London (1927).

Cla54a. Clairaut,A., "Mémoire sur l'orbite apparent du soleil autour de la *Terre*, an ayant égard aux perturbations produites par les actions de la *Lune* et des Planetes principales," *Mém. de l'Acad. des Sci. (Paris)*, pp. 521-564 (1754).

Fat28a. Fatou,P., "Sur le mouvement d'un système soumis á des forces á courte période," *Bull. Soc. Math.* **56**, pp. 98-139 (1928).

Jac69a. Jacobi,C.G.J., "Vorlesungen über Dynamik," in *Collected Works*, ed. A.Clebsch, Chelsea Publ. Co., New York (1969).

Kry37a. Krylov,N.M. and Bogoliubov,N.N., *Introduction to nonlinear mechanics (in Russian)*, Patent No. 1, Kiev, 1937.

Lag88a. Lagrange,J.-L., *Mécanique Analytique (2 vols.),* edition Albert Blanchard, Paris (1788).

Lap25a. de Laplace,S.P., *Traité de Mécanique Céleste,* Bachelier, Paris (1799-1825).

Man34a. Mandelstam,L.I. and Papalexi,N.D., "Uber die Begründung einer Methode für die Näherungslösung von Differentialgleichungen," *J. f. exp. und theor. Physik* **4**, p. 117 (1934).

Poi92a. Poincaré,H., *Les Méthodes Nouvelles de la Mécanique Céleste I,* Gauthiers-Villars, Paris (1892).

Pol26a. van der Pol,B., "On Relaxation-Oscillations," *The London, Edinburgh and Dublin Philosophical Magazine and Journal of Science* **2**, pp. 978-992 (1926).

Pol20a. van der Pol,B, "A theory of the amplitude of free and forced triode vibrations," *The Radio Review*, London **1**, pp. 701-710 (1920).

Ros66a. Roseau,M., *Vibrations nonlinéaires et théorie de la stabilité,* Springer-Verlag (1966).

Wil80a. Wilson,C.A., "Perturbations and solar tables from *Lacaille* to *Delambre*: the rapprochement of observation and theory," *Archive for History of Exact Sciences* **22**, pp. 53-188 (part I) and 189-304 (part II) (1980).

# 8.2. Multiple Time Scales Expansion

Initial value problems for equations like

$$\dot{y} = F(t,y;\epsilon)$$

lead often after some transformations to the standard form

$$\dot{x} = \epsilon f(t,x) + O(\epsilon^2)$$

and after averaging to

$$\dot{x} = \epsilon f^o(x) + O(\epsilon^2).$$

Solving the last equation to first order we obtain functions of $\tau = \epsilon t$; for the original problem in $y$ we find approximations which are functions of both $t$ and $\tau$. The variables $t$ and $\tau$ are natural variables for the problem in $y$ and this suggests the a priori introduction of such variables in the expansions. The development of this approach started in 1932 in a study by *Krylov* and *Bogoliubov* on the stability of airplanes. In 1935, *Krylov* and *Bogoliubov* presented the partial differential equations of the method together with the secularity conditions; see (Kry35a). *Kuzmak* (Kuz59a), working in the same school of mathematics, used the idea again. Independently, the method of using multiple time-like variables was developed by *Kevorkian* (Kev61a), *Cochran* (Coc62a) and *Mahony* (Mah62a). One can find modifications of the method and many examples in (Nay73a).

We shall discuss two aspects of the method. First, one finds very often in the literature claims of uniform validity of multiple time scales expansions, i.e. asymptotic validity for all time. This has not been proved and is in general not true. It has been demonstrated in (Mor66a) for the first terms of the expansion and in (Per69a) for the expansion to $O(\epsilon^n)$ that employing two time scales, $t$ and $\tau$, the method is equivalent to averaging. This implies validity of the resulting approximations on $\frac{1}{\epsilon}$. No theorems have been proved for problems with more than two time scales.

Examples have been given where on comparing multiple time scales expansions with known exact solutions the validity for all time could be established. It is interesting to note that in all such examples dissipative terms occur which give rise to attraction properties of the solutions. It has been established in chapter 4 that in such cases the averaging method leads to approximations valid for all time. The formal equivalence of the two methods means that in such cases the uniform validity carries over to the two time scales method.

A second important aspect of the method is its use in practice, the computational efficiency. Of course this is to some extent a matter of taste (and experience). However the multiple time scales method applied to initial value problems is often more laborious and far less efficient than the averaging method. In considering *Hamiltonian* perturbation problems there is the additional problem of keeping the transformations symplectic; using averaging methods one meets the same problem, but its solution is then fairly simple. In singular perturbation theory however the use of multiple time scales is sometimes more attractive.

We shall demonstrate the equivalence of averaging and the method of two time scales for systems in the standard form using (Per69a) and (Bur74a). To use the standard form is not essential but it shortens the treatment. Consider the initial value problem

$$\dot{x}=\epsilon f(t,x)\ ,\ x(0)=x_o,$$

with $x \in D \subset R^n$; $f(t,x)$ is $T$-periodic in $t$ and meets other requirements which will be formulated in the course of our calculations. Suppose that two time scales suffice for our treatment, $t$ and $\tau$, a fast and a slow time. We expand

$$x=\sum_{n=0}^{N} \epsilon^n x_n(t,\tau)$$

in which $t$ and $\tau$ are used as independent variables. The differential operator becomes

$$\frac{d}{dt} = \frac{\partial}{\partial t} + \epsilon \frac{\partial}{\partial \tau}.$$

The initial values become

$$x_o(0,0)=x_o,$$

$$x_i(0,0)=0\ ,\ i>0,$$

Substitution of the expansion in the equation yields

$$\frac{\partial x_o}{\partial t}+\epsilon\frac{\partial x_o}{\partial \tau}+\epsilon\frac{\partial x_1}{\partial t}+\cdots=\epsilon f(t,x_o+\epsilon x_1+\cdots).$$

To expand $f$ we assume $f$ to be sufficiently differentiable:

$$f(t,x_o+\epsilon x_1+\cdots)=f(t,x_o)+\epsilon\nabla f(t,x_o)\cdot x_1+\cdots.$$

Collecting terms of the same order in $\epsilon$ we have the system

$$\frac{\partial x_o}{\partial t}=0,$$

$$\frac{\partial x_1}{\partial t}=-\frac{\partial x_o}{\partial \tau}+f(t,x_o),$$

$$\frac{\partial x_2}{\partial t}=-\frac{\partial x_1}{\partial \tau}+\nabla f(t,x_o)\cdot x_1,$$

$$\cdots.$$

which can be solved successively. Integrating the first equation produces

$$x_o=A(\tau)\ ,\ A(0)=x_o.$$

At this stage $A(\tau)$ is still undetermined. Integrating the second equation produces

$$x_1=\int_0^t[-\frac{dA}{d\tau}+f(s,A(\sigma))]ds+B(\tau)\ ,\ B(0)=0,$$

$(\sigma=\epsilon s)$.

Note that the integration involves mainly $f(t,A)$ as a function of $t$, since the $\tau$-dependent contribution will be small. We wish to avoid terms in the expansion that become unbounded with time $t$. We achieve this by the *secularity condition*

$$\int_0^T[-\frac{dA}{d\tau}+f(s,A(\sigma))]ds=0\ ,\ B(\tau)\ \textit{bounded.}$$

From the integral we find

$$\frac{dA}{d\tau}=\frac{1}{T}\int_0^T f(s,A)ds\ ,\ A(0)=x_o,$$

i.e. $A(\tau)$ is determined by the same equation as in the averaging method. Then we also know from chapters 2 and 3 that

$$x(t)=A(\tau)+O(\epsilon)\ \textit{on the time-scale}\ \frac{1}{\epsilon}.$$

Note that to determine the first term $x_o$ we have to consider the expression for the second term $x_1$. This process repeats itself in the construction of higher order approximations. We abbreviate

$$x_1=u^1(t,A(\tau))+B(\tau),$$

with

$$u^1(t,A(\tau))=\int_0^t[-\frac{dA}{d\tau}+f(s,A(\sigma))]ds.$$

We find

$$x_2 = \int_0^t [-\frac{\partial x_1}{\partial \tau} + \nabla f(t,x_o)\cdot x_1] dt + C(\tau)$$

$$= \int_0^t [-\frac{\partial u^1}{\partial \tau} - \frac{dB}{d\tau} + \nabla f(t,A)\cdot u^1(t,A) + \nabla f(t,A)\cdot B] dt + C(\tau),$$

where $C(0)=0$. To obtain an expansion with terms bounded in time, we apply again a *secularity condition*

$$\int_0^T [-\frac{\partial u^1}{\partial \tau} - \frac{dB}{d\tau} + \nabla f(t,A)\cdot u^1(t,A) + \nabla f(t,A)\cdot B] dt = 0.$$

By interchanging $\nabla$ and $\int$ we find

$$\frac{dB}{d\tau} = \nabla f^o(A)\cdot B(\tau) + \frac{1}{T}\int_0^T [-\frac{\partial u^1}{\partial \tau} + \nabla f(t,A)\cdot u^1(t,A)] dt \ , \ B(0)=0,$$

where $f^o(A)$ is the average of $f(t,A)$. This is a linear inhomogeneous equation for $B(\tau)$ with variable coefficients.

The first order approximations by averaging and by the two time scales method are identical but the second order approximations are different; see § 3.5 where we obtained an $O(\epsilon^2)$ approximation on the time scale $\frac{1}{\epsilon}$ by averaging in the periodic case. To prove that the second order results are asymptotically equivalent we have a theorem; to ease the treatment we do not attempt to use the weakest assumptions possible.

**8.2.1. Theorem**

Consider the initial value problem

$$\dot{x} = \epsilon f(t,x) \ , \ x(0)=x_o,$$

with $f$, $\nabla f$ and $\nabla^2 f$ defined, continuous and bounded in $D\times[0,\infty)$, $D\subset R^n$. $A(\tau)$ and $x_1$ are constructed as indicated above. If $A(\tau)$ belongs to an interior subset of $D$ we have

$$x(t) = A(\tau) + \epsilon x_1(t, A(\tau)) + O(\epsilon^2) \textit{ on the time-scale } \frac{1}{\epsilon}.$$

**Proof**

From the construction we have $x(t)=A(\tau)+O(\epsilon)$ on the time-scale $\frac{1}{\epsilon}$. In the expression for $x_2$ we put $C(\tau)=0$. Furthermore we have from the assumptions

$$\|f\| + \|\nabla f\| + \|\nabla^2 f\| \leq M_1.$$

Moreover, cf. § 2.6,

$$\|u^1\| + \|\nabla u^1\| = O(M_1 T).$$

From the boundedness of $x_2$ which we imposed we have

$$\|x_2\| + \|\nabla x_2\| \leqslant M_2.$$

We introduce (as abbreviation)

$$y = A + \epsilon x_1 + \epsilon^2 x_2.$$

The triangle inequality yields

$$\|x(t) - A(\tau) - \epsilon x_1(t, A(\tau))\| \leqslant \|x(t) - y\| + \epsilon^2 \|x_2\| \leqslant \|x(t) - y\| + \epsilon^2 M_2.$$

We estimate $\|x(t) - y\|$ where $y = y(t, \tau)$. We compute

$$\frac{dx}{dt} - \frac{dy}{dt} = \epsilon f(t, x) - \epsilon \frac{dA}{d\tau} - \epsilon \frac{\partial x_1}{\partial t} - \epsilon^2 \frac{\partial x_2}{\partial t} - \epsilon^2 \frac{\partial x_1}{\partial \tau} - \epsilon^3 \frac{\partial x_2}{\partial \tau}$$

$$= \epsilon f(t, x) - f(t, y) + R,$$

where

$$R = \epsilon f(t, A + \epsilon x_1 + \epsilon^2 x_2) - \epsilon \frac{dA}{d\tau} - \epsilon\left(-\frac{dA}{d\tau} + f(t, A)\right)$$

$$-\epsilon^2\left[-\frac{\partial x_1}{\partial \tau} + \nabla f(t, A) \cdot x_1\right] - \epsilon^2 \frac{\partial x_1}{\partial \tau} - \epsilon^3 \frac{\partial x_2}{\partial \tau}$$

$$= \epsilon f(t, A + \epsilon x_1 + \epsilon^2 x_2) - \epsilon f(t, A) - \epsilon^2 \nabla f(t, A) \cdot x_1 - \epsilon^3 \frac{\partial x_2}{\partial \tau}.$$

Expanding $f$ with respect to $\epsilon$ we have

$$R = \epsilon^3 \nabla f(t, A) \cdot x_2 + \tfrac{1}{2} \epsilon^3 \nabla^2 f(t, A) \cdot x_1 - \epsilon^3 \frac{\partial x_2}{\partial \tau} + O(\epsilon^4).$$

The dependence of $x_2$ on $\tau$ only takes place via $A(\tau)$; we then have that $\dfrac{\partial x_2}{\partial \tau}$ is bounded. Together with the other assumptions of boundedness we find

$$\|R\| \leqslant M_3 \epsilon^3.$$

Using the *Lipschitz*-continuity of $f$ we conclude

$$\|x - y\| \leqslant \epsilon \int_0^t M_1 \|x - y\| dt + M_3 \epsilon^3.$$

The *Gronwall* Lemma 1.3.1 yields

$$\|x - y\| \leqslant \frac{M_3}{M_1} \epsilon^2 e^{M_1 \epsilon t}.$$

so that $\|x - y\| = O(\epsilon^2)$ on the time-scale $\dfrac{1}{\epsilon}$. This completes the proof. □

**References**

Bur74a. van der Burgh,A.H.P., *Studies in the Asymptotic Theory of Non-linear Resonance,* Technical Univ., Delft (1974).

Coc62a. Cochran,J.A., *Problems in singular perturbation theory,* Stanford University (1962).

Kev61a. Kevorkian,J., *The uniformly valid asymptotic representation of the solution of certain nonlinear differential equations,* Calif. Inst. Techn., Pasadena (1961).

Kry35a. Kryloff,N. and Bogoliùbov,N., "Méthodes approchées de la mécanique non linéaire dans leur application à l'étude de la perturbation des mouvements périodiques et de divers phénomènes de résonance s'y rapportant," *Académie des Sciences d'Ukraine* **14** (1935).

Kuz59a. Kuzmak,G.E., "Asymptotic solutions of nonlinear second order differential equations with variable coefficients," *J. Appl. Math. Mechanics (PMM)* **10**, pp. 730-744 (1959).

Mah62a. Mahony,J.J., "An expansion method for singular perturbation problems," *J. Australian Math. Soc.* **2**, pp. 440-463 (1962).

Mor66a. Morrison,J.A., "Comparison of the modified method of averaging and the two variable expansion procedure," *SIAM Rev.* **8**, pp. 66-85 (1966).

Nay73a. Nayfeh,A.H., *Perturbation Methods,* Wiley-Interscience, New York (1973).

Per69a. Perko,L.M., "Higher order averaging and related methods for perturbed periodic and quasi-periodic systems," *SIAM J. Appl. Math.* **17**, pp. 698-724 (1969).

# 8.3. The Forced Mathematical Pendulum

In this appendix we consider the equation

$$\ddot{\phi}+\sin(\phi)=\epsilon\sin(\omega t).$$

Although maybe not immediately obvious, one can use averaging to analyze this problem. We shall make heavy use of results on elliptic functions, all of which can be found in (Byr71a). First we consider the unperturbed problem

$$\ddot{\phi}+\sin(\phi)=0.$$

We integrate this to obtain

$$\tfrac{1}{2}\dot{\phi}^2-\cos(\phi)=\textit{constant}=c.$$

Putting $\phi=2\theta$ and $k=(\frac{2}{1+c})^{\frac{1}{2}}$, we obtain

$$\dot{\theta}^2=\frac{1}{k^2}(1-k^2\sin^2(\theta)),$$

implying

$$t=k\int_{\theta_o}^{\theta}\frac{d\phi}{(1-k^2\sin^2(\phi))^{\frac{1}{2}}}=ku,$$

which, by definition of the *Jacobian* elliptic functions *am*, *sn* and *dn*, gives

$$\sin(\theta)=sn(ku,k),$$

$$\theta=am(ku,k),$$

$$\dot{\theta}=\pm\frac{1}{k}dn(ku,k).$$

This motivates the following coordinate transformation $(\theta,\dot{\theta})\mapsto(u,k)$:

$$\theta = am(ku,k),$$

$$\dot{\theta} = \frac{1}{k} dn(ku,k),$$

which is valid in the upper half of the phase space, that is for those $(\theta,\dot{\theta})$ with $\dot{\theta}>0$.

We consider the perturbation problem

$$\ddot{\phi} + \sin(\phi) = \epsilon F(t,\phi,\dot{\phi}).$$

Using the same transformation as in the unperturbed problem, this leads to

$$\dot{\theta} = v,$$

$$\dot{v} = -\sin(\theta)\cos(\theta) + \frac{\epsilon}{2} F(t,2\theta,2\dot{\theta}).$$

Let

$$\theta = am(ku,k),$$

$$v = \frac{1}{k} dn(ku,k).$$

Using (Byr71a), formula 710, we find, with $k'^2 = 1-k^2$,

$$\frac{\partial\theta}{\partial k} = \frac{dn(ku,k)}{kk'^2}[-E(ku)+k^2 sn(ku,k)cd(ku,k)],$$

$$\frac{\partial\theta}{\partial u} = k\ dn(ku,k),$$

$$\frac{\partial v}{\partial k} = -\frac{1}{k^2} dn(ku,k) + \frac{sn(ku,k)cn(ku,k)}{k'^2}[E(ku)-dn(ku,k)tn(ku,k)],$$

$$\frac{\partial v}{\partial u} = -k^2 sn(ku,k)cn(ku,k),$$

where $E(x) = \int_0^{\frac{1}{2}\pi}(1-x^2\sin^2(\theta))^{\frac{1}{2}} d\theta$ and the *Jacobian* determinant of the transformation is given by

$$\Delta = \frac{\partial\theta}{\partial k}\frac{\partial v}{\partial u} - \frac{\partial\theta}{\partial u}\frac{\partial v}{\partial k} = \frac{1}{k},$$

which follows from the definitions of the *Jacobian* elliptic functions.

With this information, we can easily compute the vectorfield in the new variables:

$$\dot{k} = \frac{1}{\Delta}(\frac{\partial v}{\partial u}\dot{\theta} - \frac{\partial\theta}{\partial u}\dot{v}) =$$

$$-\frac{\epsilon}{2}k^2 dn(ku,k)F(t,2am(ku,k),\frac{2}{k}dn(ku,k)),$$

$$\dot{u} = -\frac{1}{\Delta}(\frac{\partial v}{\partial k}\dot{\theta} - \frac{\partial\theta}{\partial k}\dot{v} =$$

$$= \frac{1}{k^2} + \tfrac{1}{2}\epsilon \frac{1}{k'^2}(-E(ku)dn(ku,k) + k^2 sn(ku,k)cn(ku,k))F.$$

### 8.3.1. Remark

On first sight there seems to be a singularity in this equation for $k$ approaching 1, or $k'$ small. Using, however, formula 127.02 in (Byr71a), we find that

$$sn(u) = \tanh(u) + O(k'^2),$$

$$cn(u) = sech(u) + O(k'^2),$$

$$dn(u) = sech(u) + O(k'^2),$$

$$E(ku) = \int_0^{ku} dn^2(w)dw \sim \int_0^{u} sech^2(w)dw = \tanh(u),$$

which implies that $\dot{u} = \frac{1}{k^2} + O(\epsilon)$, uniform in $k$.

We shall now specify the problem to the case $F = \sin(\omega t)$. Using formula 908.03 in (Byr71a), we find

$$dn(ku,k) = \frac{\pi}{2K(k)} + \frac{2\pi}{K(k)} \sum_{m=0}^{\infty} \frac{q^{m+1}}{(1+q^{2(m+1)})} \cos(m+1)\frac{\pi ku}{K(k)},$$

where

$$K(x) = \int_0^{\frac{1}{2}\pi} \frac{d\theta}{(1-x^2\sin^2(\theta))^{\frac{1}{2}}}$$

and

$$q = e^{-\frac{\pi K'(k)}{K(k)}}$$

or

$$dn(ku,k) = \frac{\pi}{2K(k)} \sum_{-\infty}^{\infty} \frac{\cos(\frac{m\pi ku}{K(k)})}{\cosh(\frac{m\pi K'(k)}{K(k)})}.$$

This leads to

$$\dot{k} = \epsilon \frac{k^2}{4} \frac{\pi}{K(k)} \sum_{-\infty}^{\infty} \frac{1}{\cosh(\frac{m\pi K'}{K(k)})} \sin(\frac{m\pi ku}{K} - \omega t),$$

$$\dot{u} = \frac{1}{k^2} + O(\epsilon).$$

Except for the fact that we have infinitely many terms, this is of the type studied in chapter 5. We shall construct the averaging transformation and analyze the convergence of the formal series.

We have a system of type

$$\dot{x}=\epsilon X(\phi,x)\ ,\ x\in\mathbf{R},$$
$$\dot{\phi}=\Omega(x)+O(\epsilon)\ ,\ \phi\in\mathbf{T}^2,$$

with $x=k$, $\phi_1=u$ and $\phi_2=t$.

$$X(\phi,x)=\frac{x^2\pi}{4K(x)}\sum_{-\infty}^{\infty}\frac{1}{\cosh(\frac{m\pi K'}{K})}\sin(\frac{m\pi x\phi_1}{K}-\phi_2),$$

$$\Omega(x)=\begin{pmatrix}\frac{1}{x^2}\\ \omega\end{pmatrix}.$$

We define $\phi_m=\dfrac{m\pi x\phi_1}{K(x)}-\phi_2$ and this leads to

$$\dot{x}=\epsilon\sum_{-\infty}^{\infty}X^m(\phi_m,x),$$

$$\dot{\phi}_m=\frac{m\pi}{xK(x)}-\omega+O(\epsilon)=\Omega_m(x)+O(\epsilon),$$

with

$$X^m(\phi_m,x)=\frac{x^2\pi}{4K(x)}\frac{1}{\cosh(\frac{m\pi K'}{K})}\sin(\phi_m).$$

As usual, we let

$$U^{\hat{r}}(\phi,x)=\sum_{m\neq r}U^m(\phi_m,x),$$

with

$$U^m(\phi_m,x)=\frac{\pi x^2}{4K(x)\Omega_m(x)}\frac{1}{\cosh(\frac{m\pi K'(x)}{K(x)})}\cos(\phi_m),$$

where $r$ (for resonant) is such that $\Omega_r(x)$ is small (compared to $\Omega_{r\pm 1}(x)$) and $r>0$.

We try to find an averaging transformation valid outside some resonance region. The difficulty is that $r$ is a function of $x$. If the formal sequence

$$\sum_{m\neq r}U^m(\phi_m,x)$$

converges, then this defines $U^{\hat{r}}$ and one can differentiate this expression to show that the transformation

$$x=y+\epsilon U^{\hat{r}}(\phi,y)$$

does average the equation outside the resonance domain. We find, since

$$\Omega_m(x)=\frac{m\pi}{xK(x)}-\omega\sim\frac{(m-r)\pi}{xK(x)},$$

$$|\hat{U^r}(\phi,x)|\leqslant\sum_{m\neq r}|\frac{xK(x)}{(m-r)\pi}\frac{\pi}{4}\frac{x^2}{K(x)}\frac{1}{\cosh(\frac{m\pi K'(x)}{K(x)})}\cos(\phi_m)|\leqslant$$

$$\leqslant\frac{x^3}{2}e^{\frac{r\pi K'(x)}{K(x)}}\sum_{m=1}^{\infty}\frac{1}{m}e^{-m\pi\frac{K'(x)}{K(x)}}$$

$$=\frac{x^3}{2}e^{\frac{r\pi K'(x)}{K(x)}}\int_{\frac{\pi K'(x)}{K(x)}}^{\infty}\frac{e^{-y}}{1-e^{-y}}dy$$

$$=\frac{x^3}{2}e^{\frac{r\pi K'(x)}{K(x)}}\log|1-e^{-\frac{\pi K'(x)}{K(x)}}|.$$

We shall now consider the following limit: we let $x$ go to 1 such that $\Omega_r(x)$ stays small; this we call the *resonant limit.* In this limit, we have

$$\frac{r\pi}{xK(x)}\sim\omega\ , \quad so\ r\to\infty,\ but\ \frac{r}{K(x)}\sim\frac{\omega}{\pi}.$$

Furthermore, $K'(1)=\frac{1}{2}\pi$ and we find

$$res.\ lim\ \frac{x^3}{2}e^{\frac{r\pi K'(x)}{K(x)}}=\frac{1}{2}e^{\frac{1}{2}\omega\pi}.$$

The expression $\log|1-e^{-\frac{\pi K'(x)}{K(x)}}|$ however behaves as follows:

$$\log|1-e^{-\frac{\pi K'(x)}{K(x)}}|\sim\log(\frac{\pi}{K(x)}\sim\log(r).$$

So we have

$$\hat{U^r}=O(\epsilon\log(r)).$$

Having obtained an estimate for the averaging transformation, which indicates some trouble near $k=1$, we take the resonant limit in the averaged equation, given by

$$\dot{x}=\epsilon\frac{\pi}{4}\frac{x^2}{K(x)}\frac{1}{\cosh(\frac{r\pi K'(x)}{K(x)})}\sin(\phi_r).$$

In the resonant limit, this equals

$$\dot{x}=\epsilon\frac{\pi}{4}\frac{1}{K(x)}\frac{1}{\cosh(\frac{1}{2}\pi\omega)}\sin(\phi_\infty),$$

where the exact meaning of $\phi_\infty$ remains to be determined. One should remark the likeness of this expression with the *Melnikov*-function of this problem (Cf. appendix 8.4):

$$\Delta_o = -\frac{2\pi \sin(c)}{\cosh(\frac{1}{2}\pi\omega)}$$

where $c$ parametrizes the separatrix (and could be identified to $\phi_\infty$).

## References

Byr71a. Byrd,P.F. and Friedman,M.B., *Handbook of Elliptic Integrals for Engineers and Scientists,* Springer Verlag (1971).

# 8.4. The Melnikov Function

We shall give here a formal derivation of the *Melnikov* function (Mel63a) (See also (Hol80a)) for planar systems. The error estimates, using the invariant manifold theorem and 'shadowing', can be found in (San82a), along with further references to the literature.

We consider a system of the type

$$\dot{x}=f^{o}(x)+\epsilon f^{1}(t,x;\epsilon)\ ,\ x\in D\subset\mathbf{R}^{2}.$$

Let $x_o$ be a homoclinic point of the unperturbed ($\epsilon=0$) equation (everything works fine with heteroclinic connections too). We denote any solution of the unperturbed system with $x_o$ as its limit set by $x_o^{s,u}$ (s and u for stable and unstable). From the invariant manifold theorem (Cf. (Hir77a)) we know that there exist solutions $x_\epsilon^s$ and $x_\epsilon^u$ of the perturbed system with the following limiting behavior:

$$\lim_{t\to\infty}x_\epsilon^s(t)=x_o^\epsilon\ ,\ \lim_{t\to-\infty}x_\epsilon^u(t)=x_o^\epsilon,$$

with $x_o^\epsilon=x_o+O(\epsilon)$. We let

$$x_\epsilon^i(t)=x_o^{s,u}(t)+\epsilon\xi^i(t;\epsilon)\ ,\ i=s,u.$$

The functions $\xi^i$ are not uniformly bounded, but one can derive exponential estimates. Let

$$x=x_o^{s,u}(0)$$

and define

$$\Delta_\epsilon^i(t,x)=f^o(x_o^{s,u}(t))\wedge\xi^i(t;\epsilon)\ ,\ i=s,u$$

(where $x\wedge y$ denotes the area spanned by the planar vectors $x$ and $y$). We can write down a differential equation for $\Delta_\epsilon^i$ as follows:

$$\dot{\Delta}_\epsilon^i = trace \nabla f^o(x_o^{s,u}(t)) \cdot \Delta_\epsilon^i + f^o(x_o^{s,u}(t)) \wedge f^1(t, x_o^{s,u}(t);0) + \cdots .$$

Since

$$\lim_{t\to\infty} \Delta_\epsilon^s(t,x) = \lim_{t\to-\infty} \Delta_\epsilon^u(t,x) = 0$$

we can integrate this differential equation as follows:

$$\Delta_\epsilon^s(t,x) = -\int_t^\infty e^{-\int_t^\tau trace \nabla f^o(x_o^{s,u}(\sigma))d\sigma} f^o(x_o^{s,u}(\tau)) \wedge f^1(\tau, x_o^{s,u}(\tau);0)d\tau,$$

$$\Delta_\epsilon^u(t,x) = -\int_{-\infty}^t e^{\int_t^\tau trace \nabla f^o(x_o^{s,u}(\sigma))d\sigma} f^o(x_o^{s,u}(\tau)) \wedge f^1(\tau, x_o^{s,u}(\tau);0)d\tau$$

(where we neglect the $O(\epsilon)$-terms).

We define the *Melnikov* function $\Delta_o$ as

$$\Delta_o(t,x) = \Delta_o^u(t,x) - \Delta_o^s(t,x)$$

$$= \int_{-\infty}^\infty e^{\int_t^\tau trace \nabla f^o(x_o^{s,u}(\sigma))d\sigma} f^o(x_o^{s,u}(\tau)) \wedge f^1(\tau, x_o^{s,u}(\tau);0)d\tau.$$

In many problems, $trace \nabla f^o = 0$, and the *Melnikov* function simplifies to:

$$\Delta_o(x) = \int_{-\infty}^\infty f^o(x_o^{s,u}(\tau)) \wedge f^1(\tau, x_o^{s,u}(\tau);0)d\tau.$$

In general, $\Delta_o$ will be an approximation to $\Delta_\epsilon$ (defined as $\Delta_\epsilon^u - \Delta_\epsilon^s$), and, using the implicit function theorem, we can find the zeros of $\Delta_\epsilon$ from $\Delta_o$ under transversality conditions. Since intersection of the stable and unstable manifold makes $\Delta_\epsilon$ vanish by definition, we know that $|\Delta_o| > 0$ implies that there can be no intersection of the stable and unstable manifold.

### 8.4.1. Example: the forced pendulum

Consider

$$\ddot{\phi} + \sin(\phi) = \epsilon \sin(\omega t).$$

We write this as

$$\dot{\phi} = x,$$

$$\dot{x} = -\sin(\phi) + \epsilon \sin(\omega t).$$

Then

$$f^o(\phi,x) = \begin{bmatrix} x \\ -\sin(\phi) \end{bmatrix}, \quad f^1(t,\phi,x) = \begin{bmatrix} 0 \\ \sin(\omega t) \end{bmatrix},$$

$$f^o \wedge f^1 = x \sin(\omega t),$$

$$x_o^{s,u}(t) = \frac{2}{\cosh(t+c)} \quad , \; c = \log(\tan(\frac{\pi}{4} + \frac{\phi_o}{4}))$$

(where $\phi_o$ is the initial condition for $\phi_o^{s,u}$). Evaluating $\Delta_o$ leads to

$$\Delta_o(\phi_o) = -\frac{2\pi\sin(\omega c)}{\cosh(\frac{\omega\pi}{2})}.$$

For an application of the *Melnikov* function to the *Josephson* equation

$$\beta\ddot{\phi} + (1 + \gamma\cos(\phi))\dot{\phi} + \sin(\phi) = \alpha + k\omega\sin(\omega t),$$

see (San82a), where further references to applications can also be found.

**References**

Hir77a. Hirsch,M.W., Pugh,C.C., and Shub,M., "Invariant Manifolds," *Springer Lecture Notes in Mathematics*, Berlin **583**, Springer-Verlag (1977).

Hol80a. Holmes,P.J., "Averaging and chaotic motions in forced oscillations," *SIAM J.Appl.Math.* **38**, pp. 65-80 (1980).

Mel63a. Melnikov,V.K., "On the stability of the center for time periodic perturbations," *Trans.Moscow Math.Soc.* **12**(1), pp. 1-57 (1963).

San82a. Sanders,J.A., "Melnikov's method and averaging," *Celestial Mechanics* **28**, pp. 171-181 (1982).

# 8.5. The 1:2:3-Resonance

In action-angle coordinates, the normal form of the exact 1:2:3-resonance (detuning parameters both zero) is given by

$$H=\tau_1+2\tau_2+3\tau_3+2\epsilon(2\tau_1\tau_2)^{1/2}[a_1\tau_1^{1/2}\cos(2\phi_1-\phi_2-\alpha_1)$$
$$+a_2\tau_3^{1/2}\cos(\phi_1+\phi_2-\phi_3-\alpha_2)].$$

Introducing the combination angles

$$\psi_1=\phi_1+\phi_2-\phi_3-\alpha_1,$$
$$\psi_2=2\phi_1-\phi_2-\alpha_2,$$

we find the equations of motion

$$\dot{\tau}_1=2\epsilon(2\tau_1\tau_2)^{1/2}[a_1\tau_3^{1/2}\sin(\psi_1)+2a_2\tau_1^{1/2}\sin(\psi_2)],$$
$$\dot{\tau}_2=2\epsilon(2\tau_1\tau_2)^{1/2}[a_1\tau_3^{1/2}\sin(\psi_1)-a_2\tau_1^{1/2}\sin(\psi_2)],$$
$$\dot{\tau}_3=-2\epsilon a_1(2\tau_1\tau_2\tau_3)^{1/2}\sin(\psi_1),$$
$$\dot{\psi}_1=2\epsilon(2\tau_1\tau_2\tau_3)^{-1/2}[a_1(\tau_1\tau_3+\tau_2\tau_3-\tau_1\tau_2)\cos(\psi_1)$$
$$+a_2(\tau_1\tau_3)^{1/2}(\tau_1+2\tau_2)\cos(\psi_2)],$$
$$\dot{\psi}_2=\epsilon(\tau_1\tau_2)^{-1/2}[a_1(2\tau_3)^{1/2}(2\tau_2-\tau_1)\cos(\psi_1)+a_2(2\tau_1)^{1/2}(4\tau_2-\tau_1)\cos(\psi_2)].$$

The new coordinates are not canonical. Of course, we could have lifted the transformation of the angles to the actions, but the expressions for the new actions are rather ugly. We find the periodic orbits as follows: We assume $\tau_1$ , $\tau_2$ and $\tau_3$ to be strictly positive. Then $\dot{\tau}_3=0$ implies

$$\sin(\psi_2)=0$$

and $\dot{\tau}_2=0$ implies

$$\sin(\psi_1)=0.$$

Thus $\cos(\psi_1)=\pm 1$ and $\cos(\psi_2)=\pm 1$. Let $\bar{a}_1=a_1\cos(\psi_1)$ and $\bar{a}_2=a_2\cos(\psi_2)$. Then

$$\bar{a}_1(2\tau_1)^{1/2}(4\tau_2-\tau_1)+\bar{a}_2(2\tau_3)^{1/2}(2\tau_2-\tau_1)=0, \qquad \text{A5-1}$$

$$\bar{a}_1(\tau_1\tau_3)^{1/2}(\tau_1+2\tau_2)+\bar{a}_2(\tau_1\tau_3+\tau_2\tau_3-\tau_1\tau_2)=0. \qquad \text{A5-2}$$

Multiplying (A5-1) with $\tau_3^{1/2}(\tau_1+2\tau_2)$ and (A5-2) with $\tau_2^{1/2}(4\tau_2-\tau_1)$, we find

$$\tau_1-4\tau_2+3\tau_3=0. \qquad \text{A5-3}$$

Combined with

$$\tau_1+2\tau_2+3\tau_3=E$$

this results in

$$\tau_2=\frac{E}{6}.$$

Substituting (A5-3) in (A5-1), we find

$$3\bar{a}_1(2\tau_1)^{1/2}\tau_3+\bar{a}_2(2\tau_3)^{1/2}(2\tau_2-\tau_1)=0$$

or, again substituting (A5-3),

$$6\bar{a}_1(\tau_1\tau_3)^{1/2}+\bar{a}_2(3\tau_3-\tau_1)=0.$$

Solving this quadratic equation (in $\tau_1^{1/2}$ and $\tau_3^{1/2}$), we find

$$\left(\frac{\tau_1}{\tau_3}\right)^{1/2}=3\frac{\bar{a}_1}{\bar{a}_2}\pm\left(9\left(\frac{\bar{a}_1}{\bar{a}_2}\right)^2+3\right)^{1/2}.$$

The minus sign would lead to negative values for the square root, so we consider only the + sign. Written in the original variables, this is

$$\left(\frac{\tau_1}{\tau_3}\right)^{1/2}=\left(9\left(\frac{a_1}{a_2}\right)^2+3\right)^{1/2}\pm 3\frac{a_1}{a_2}.$$

Let $\beta=3\dfrac{\bar{a}_1}{\bar{a}_2}\pm\left(9\left(\dfrac{\bar{a}_1}{\bar{a}_2}\right)^2+3\right)^{1/2}$. Then

$$\tau_1+3\tau_3=\frac{2}{3}E,$$

$$\tau_1=\beta^2\tau_3$$

or

$$\tau_3=\frac{2}{3}\frac{E}{\beta^2+3}.$$

So the periodic orbit is given by

$$\cos(\psi_1^o)=\pm 1,$$

$$\cos(\psi_2^o) = \pm 1,$$

$$\beta = 3\frac{a_1\cos(\psi_1^o)}{a_2\cos(\psi_2^o)} + (9(\frac{a_1}{a_2})^2 + 3)^{1/2},$$

$$\tau_1^o = \frac{2}{3}\frac{E\beta^2}{\beta^2+3},$$

$$\tau_2^o = \frac{E}{6},$$

$$\tau_3^o = \frac{2}{3}\frac{E}{\beta^2+3}.$$

This gives four periodic orbits, characterized by their $(\psi_1,\psi_2)$-values, with (0,0) and $(\pi,\pi)$ on the same action level, as well as $(0,\pi)$ and $(\pi,0)$. If we linearize the vectorfield at the equilibria, using the translation

$$\tau_i = \tau_i^o + R_i,$$

$$\psi_i = \psi_i^o + \Psi_i$$

we find, eliminating $R_2$ with $R_1 + 2R_2 + 3R_3 = 0$, $\tau_2^o = \frac{1}{4}(\tau_1^o + 3\tau_3^o)$ and $\bar{a}_1$ with $\bar{a}_1 = -\frac{\bar{a}_2}{6(\tau_1^o\tau_3^o)^{1/2}}(3\tau_3^o - \tau_1^o)$, the linear system

$$\frac{d}{dt}\begin{pmatrix} R \\ \Psi \end{pmatrix} = \begin{pmatrix} 0 & A \\ B & 0 \end{pmatrix}\begin{pmatrix} R \\ \Psi \end{pmatrix},$$

where $R = \begin{pmatrix} R_1 \\ R_3 \end{pmatrix}$, $\Psi = \begin{pmatrix} \psi_1 \\ \psi_2 \end{pmatrix}$ and $A$ and $B$ are $2\times 2$-matrices. The eigenvalues of this system are determined by

$$\lambda^4 - Tr(AB)\lambda^2 + \det(AB) = 0,$$

where, dropping the $^o$-index:

$$Tr(AB) = -\frac{2}{3}\epsilon^2 a_2^2 \frac{\tau_2}{\tau_1\tau_3}(7\tau_1^2 + 8\tau_1\tau_3 - 3\tau_3^2),$$

$$Det(AB) = \frac{64}{3}\epsilon^4 a_2^4 \tau_2^3 \tau_3^{-2}(\tau_1 - 3\tau_3).$$

For a further analysis of this resonance we refer to (Aa,83a), where all details are given.

## References

Aa,83a. van der Aa,E., "First order resonances in three-degrees-of-freedom systems," *Celestial Mechanics* **31**, pp. 163-191 (1983).

# 8.6. An Example of Hopf Bifurcation

## 8.6.1. Introduction

We present here the essentials of the *Hopf*-bifurcation theory, as far as they might be of use to the actual user, and, on the other hand, we boil down the amount of computations needed, to the point where they will not present the reason for not computing anything at all.
There are many computational techniques and it is difficult to make a choice, not only for the engineer who wants to apply all these ideas to some real life problem, but also for the mathematician, who wants to know what 'theorems' can in fact be proven about some asymptotic or numerical approximation. It has been one of our goals to make life a bit easier for both kind of people; we do not believe that practical computability and provability are contradictory requirements on a theory, and certainly not here. On the contrary, it often proves easier to prove something when the computations involved are easy and systematic, than to do the same thing in a method requiring a lot of experience and understanding of the problem, like the method of multiple-time scales.

The actual problem to be treated here as the model problem for the application of our techniques, was partly solved in (Set77a). For two values of the parameter the asymptotic computation was carried out and (successfully) compared to the numerical result obtained separately. Since the asymptotic computations were only done for numerical values of the parameters, no general formula for the bifurcation behavior was obtained by these authors. In this appendix we shall derive such a formula, and we shall also be able to give the approximating solutions, derived by the method of averaging.

One of the nice things of the method of averaging is, that we have at our disposal a rather strong result on the validity of the approximations obtained. This has been described in chapter 4, and we shall not give any details here. It should be noted, however, that one needs in fact a slight generalization of this result, since here we are dealing with two time-scales on which attraction occurs. This is easy to do, when one is familiar with the theory of extension of time-scales, but rather complicated by the sheer mass of detail, when one is not.

We shall obtain the following asymptotic results: suppose that a pair of eigenvalues is very nearly purely imaginary, then we use the method of averaging to obtain $O(\epsilon)$ -approximations with validity on the time-scale $0 \leqslant \epsilon t \leqslant L$ for all components of the solution, which is the usual result, but also with validity on $[0,\infty)$ for all components but the angular. The term angular refers to the change to polar coordinates in the stable manifold, that is the plane to which all orbits are attracted.

## 8.6.2. The model problem

We take our model problem describing a follower-force system from (Set77a) where we refer the reader to for details and explanation; the equations are

$$(m_1+m_2)l^2\ddot{\phi}_1+m_2l^2\ddot{\phi}_2 cos(\phi_2-\phi_1)-m_2l^2\dot{\phi}_2^2 sin(\phi_2-\phi_1)$$

$$+2d\dot{\phi}_1-d\dot{\phi}_2+2c\phi_1-c\phi_2 = -Plsin(\phi_2-\phi_1), \qquad \text{A6-1}$$

$$m_2l^2\ddot{\phi}_2+m_2l^2\ddot{\phi}_1 cos(\phi_2-\phi_1)+m_2l^2 sin(\phi_2-\phi_1)-d\dot{\phi}_1+d\dot{\phi}_2-c\phi_1+c\phi_2 = 0.$$

Let

$$\tau = (c/m)^{1/2} t/l,$$

$$B = d/l(m/c)^{1/2},$$

$$\theta = Plg/c,$$

$$\mu = m_1/m_2$$

and scale

$$q_i = (\epsilon)^{1/2}\,\phi_i, \qquad i = 1,2,$$

where $\epsilon$ is a small, positive parameter.

Then the system (A6-1) can be written in vectorform as follows:

$$\mathbf{A}q''+\mathbf{B}q'+\mathbf{C}q = \epsilon g(q,q')+O(\epsilon^2),$$

where

$$q = \begin{pmatrix} q_1 \\ q_2 \end{pmatrix}, \quad ' = \frac{d}{d\tau}$$

and, if we take $\mu = 2$ ,

$$\mathbf{A} = \begin{pmatrix} 3 & 1 \\ 1 & 1 \end{pmatrix}, \quad \mathbf{B} = B\begin{pmatrix} 2 & -1 \\ -1 & 1 \end{pmatrix}, \mathbf{C} = \begin{pmatrix} 2-\theta & \theta-1 \\ -1 & 1 \end{pmatrix},$$

$$g = \begin{pmatrix} g_1 \\ g_2 \end{pmatrix},$$

with

$$g_1 = ¼(q_2-q_1)^2[5Bp_1-4Bp_2-(\theta-5)q_1-(4-\theta)q_2]+(q_2-q_1)(p_1)^2+\frac{\theta}{6}(q_2-q_1)^3,$$

$$g_2 = ¼(q_2-q_1)^2[-3Bp_1+2Bp_2+(\theta-3)q_1+(2-\theta)q_2]-(q_2-q_1)p_1^2$$

(Formula { 57 } in (Set77a) does contain two printing errors: in $g_1$ the cube was written as a square, and in $g_2$ the factor 1/4 has been omitted).

In the next section we will write the equation as a first order system, and after some simplifying transformations , compute the eigenvalues and -spaces of its linear part.

## 8.6.3. The linear equation

Consider the equation

$$\mathbf{A}q''+\mathbf{B}q'+\mathbf{C}q = 0.$$

Let $p = q'$ , then

$$\frac{d}{d\tau}\begin{pmatrix} q \\ p \end{pmatrix} = \begin{pmatrix} 0 & 1 \\ -\mathbf{A}^{-1}\mathbf{C} & -\mathbf{A}^{-1}\mathbf{B} \end{pmatrix}\begin{pmatrix} q \\ p \end{pmatrix},$$

provided, of course, that **A** is invertible, as is the case in our problem, Let

$$\overline{q} = \mathbf{S}q, \qquad \overline{p} = \mathbf{S}p,$$

where

$$\mathbf{S} = \begin{pmatrix} 1 & 1 \\ -1 & 1 \end{pmatrix}.$$

Then

$$\frac{d}{d\tau}\begin{pmatrix} \overline{q} \\ \overline{p} \end{pmatrix} = \begin{pmatrix} 0 & 1 \\ -\mathbf{SA}^{-1}\mathbf{CS}^{-1} & -\mathbf{SA}^{-1}\mathbf{BS}^{-1} \end{pmatrix}\begin{pmatrix} \overline{q} \\ \overline{p} \end{pmatrix}$$

or

$$\frac{d}{d\tau}\begin{pmatrix} \overline{q}_1 \\ \overline{q}_2 \\ \overline{p}_1 \\ \overline{p}_2 \end{pmatrix} = \begin{pmatrix} 0 & 0 & 1 & 0 \\ 0 & 0 & 0 & 1 \\ 0 & -1 & 0 & -B \\ \frac{1}{2} & \theta-\frac{7}{2} & \frac{B}{2} & -\frac{7}{2}B \end{pmatrix}\begin{pmatrix} \overline{q}_1 \\ \overline{q}_2 \\ \overline{p}_1 \\ \overline{p}_2 \end{pmatrix} =: A\begin{pmatrix} \overline{q} \\ \overline{p} \end{pmatrix}.$$

The characteristic equation of A is:

$$\lambda^4 + \frac{7}{2}B\lambda^3 - (\theta - \frac{7}{2} - \frac{1}{2}B^2)\lambda^2 + B\lambda + \frac{1}{2} = 0.$$

The Routh-Hurwitz criteria (for stability ) are:

$$0 < D_1 = \frac{7}{2}B,$$

$$0 < \frac{(7}{4} - (\theta - \theta_{cr}))\frac{7}{2}B = D_2 \text{ , with } \theta_{cr} = \frac{1}{2}B^2 + \frac{41}{28},$$

$$0 < D_3 = -\frac{7}{2}B^2(\theta - \theta_{cr}).$$

If $B>0$ and $\theta<\theta_{cr}$ , 0 is asymptotically stable. We are interested in the situation where $\delta = \theta - \theta_{cr}$ is small, say $O(\epsilon)$ . At $\theta = \theta_{cr}$ , we find that the equation splits as follows:

$$(\lambda^2 + 2/7)(\lambda^2 + 7B/2\,\lambda + 7/4) = 0$$

and we see that two conjugate eigenvalues are crossing the imaginary axis, while the other two still have strictly negative real parts.

We will show that it suffices to analyze the eigenspaces at $\theta = \theta_{cr}$ in order to compute the eigenvalues and -spaces with $O(\eta)$ - accuracy.

**Exercise**

Determine the $O(\eta)$-accuracy explicitly.

### 8.6.4. Linear perturbation theory

We split A as follows:

$$A = A_o + \delta A_p.$$

Suppose we found a transformation T such that

$$T^{-1}A_oT = \begin{bmatrix} \Lambda_1 & 0 \\ 0 & \Lambda_2 \end{bmatrix}$$

and such that the eigenvalues of $\Lambda_1$ are on the imaginary axis and the eigenvalues of $\Lambda_2$ on the left. Define $A_{i,j}, i,j = 1,2$ by

$$\begin{bmatrix} A_{11} & A_{12} \\ A_{21} & A_{22} \end{bmatrix} = T^{-1}A_pT, \qquad A_{ij} \in M(2,\mathbf{R}).$$

We define a near-identity transform U:

$$U = I + \delta \begin{bmatrix} 0 & X \\ Y & 0 \end{bmatrix}.$$

Then

$$U^{-1} = I - \delta \begin{bmatrix} 0 & X \\ Y & 0 \end{bmatrix} + O(\delta^2)$$

and

$$U^{-1}T^{-1}ATU = \begin{bmatrix} \Lambda_1 & 0 \\ 0 & \Lambda_2 \end{bmatrix} + \delta \begin{bmatrix} A_{11} & A_{12} + \Lambda_1 X - X\Lambda_2 \\ A_{12} + \Lambda_2 Y - Y\Lambda_1 & A_{22} \end{bmatrix} + O(\delta^2).$$

Since $\Lambda_1$ and $\Lambda_2$ have no common eigenvalues, it is possible to solve the equations

$$A_{12} = X\Lambda_2 - \Lambda_1 X, \; A_{21} = Y\Lambda_1 - \Lambda_2 Y$$

(cf, e.g. (Bel70a)) and we obtain

$$U^{-1}T^{-1}ATU = \begin{bmatrix} \Lambda_1 + \delta A_{11} & 0 \\ 0 & \Lambda_2 + \delta A_{22} \end{bmatrix} + O(\delta^2).$$

It is not necessary to compute $T^{-1}$: it suffices to know only one block; this is due to the simple form of $A_p$, in the computation of $A_{11}$. The reader will find no difficulties in following this remark, but since we did compute $T^{-1}$ anyway, we shall follow the straightforward route without thinking.

We can take T as follows

$$T = \begin{bmatrix} 1 & B & \frac{4}{7} & -B \\ \frac{2}{7} & 0 & 1 & 0 \\ -\frac{2}{7}B & 1 & -B & 1 \\ 0 & \frac{2}{7} & -\frac{7}{2}B & \frac{7}{4} \end{bmatrix}$$

and then

$$T^{-1} = \frac{-1}{\Delta} \begin{bmatrix} -\frac{41}{28} & \frac{41}{49} - \frac{139}{28}B^2 & \frac{57}{28}B & -2B \\ -B & B(\frac{57}{28} - B^2) & -(\frac{41}{28} + B^2) & \frac{41}{49} \\ \frac{41}{98} & -\frac{1}{14}(\frac{41}{2} + \frac{57}{7}B^2) & -\frac{57}{98}B & \frac{4}{7}B \\ B & B(\frac{82}{7^3} - \frac{7}{2} - B^2) & \frac{82}{7^3} - B^2 & -\frac{41}{49} \end{bmatrix}$$

where

$$\Delta = 2B^2 + \frac{41^2}{4 \cdot 7^3}.$$

It follows that

$$T^{-1}A^oT = \begin{bmatrix} 0 & 1 & 0 & 0 \\ -\frac{2}{7} & 0 & 0 & 0 \\ 0 & 0 & -\frac{7}{2}B & \frac{7}{4} \\ 0 & 0 & -1 & 0 \end{bmatrix},$$

ie

$$\Lambda_1 = \begin{bmatrix} 0 & 1 \\ -\frac{2}{7} & 0 \end{bmatrix} ; \Lambda_2 = \begin{bmatrix} -\frac{2}{7}B & \frac{7}{4} \\ -1 & 0 \end{bmatrix}$$

and

$$A_{11} = \frac{2}{\Delta} \begin{bmatrix} \frac{2}{7}B & 0 \\ -\frac{41}{7^3} & 0 \end{bmatrix}.$$

### 8.6.5. The nonlinear problem and the averaged equations

After the transformations in the linear system, we obtain nonlinear equations of the form

$$\frac{d}{d\tau}\begin{bmatrix} y \\ z \end{bmatrix} = \begin{bmatrix} \Lambda_1 & 0 \\ 0 & \Lambda_2 \end{bmatrix}\begin{bmatrix} y \\ z \end{bmatrix} + \delta\begin{bmatrix} A_{11} & 0 \\ 0 & A_{22} \end{bmatrix}\begin{bmatrix} y \\ z \end{bmatrix} + \epsilon g^*(y,z) + O((\epsilon+\delta)^2),$$

with

$$g^*(y,z) = T^{-1}\begin{bmatrix} 0 \\ \mathbf{S}g(\mathbf{S}^{-1}T_{11}y + \mathbf{S}^{-1}T_{12}z, \mathbf{S}^{-1}T_{21}y + \mathbf{S}^{-1}T_{22}z) \end{bmatrix},$$

where $T_{ij}$ are the $2\times2$ -blocks of T.

The idea is now to average over the action induced by $\exp\Lambda_1 t$ on the vectorfield and to get rid of the z-coordinate, since it is exponentially decreasing and does not influence the system in the first order approximation.

We refer to (Cho77a) for details.

The easiest way to see what is going on, is to transform y to polar coordinates:

$$y_1 = r\sin\omega\phi,$$

$$y_2 = \omega r\cos\omega\phi,$$

where $\omega^2 = \frac{2}{7}$.

The unperturbed equation ( $\epsilon = \delta = 0$ ) transforms to

$$\dot{r} = 0,$$

$$\dot{\phi} = 1,$$

$$\dot{z} = \Lambda_2 z,$$

The perturbed equations are a special case of the following type:

$$\dot{\phi} = 1+\delta\sum_{j}\tilde{X}^{j}(r,\phi)+\epsilon\sum_{\alpha,j}X^{j}_{\alpha}(r,\phi)z^{\alpha} + O((\epsilon+\delta)^2),$$

$$\dot{r} = \delta\sum_{j}\tilde{Y}^{j}(r,\phi)+\epsilon\sum_{\alpha,j}Y^{j}_{\alpha}(r,\phi)z^{\alpha} + O((\epsilon+\delta)^2),$$

$$\dot{z} = \Lambda_2 z+\delta A_{22}z+\epsilon\sum_{\alpha,j}Z^{j}_{\alpha}(r,\phi)z^{\alpha}+O((\epsilon+\delta)^2).$$

Let

$$\mathbf{X}^{j}_{\alpha} = \begin{pmatrix} X^{j}_{\alpha} \\ Y^{j}_{\alpha} \\ Z^{j}_{\alpha} \end{pmatrix} \quad \text{and} \quad \tilde{\mathbf{X}}^{j} = \begin{pmatrix} \tilde{X}^{j} \\ \tilde{Y}^{j} \end{pmatrix}.$$

$\mathbf{X}$ and $\tilde{\mathbf{X}}$ are defined by

$$\frac{\partial^2}{\partial\phi^2}\mathbf{X}^{j}_{\alpha}+\omega^2 j^2\mathbf{X}^{j}_{\alpha} = 0, \qquad \frac{\partial^2}{\partial\phi^2}\tilde{\mathbf{X}}^{j} + \omega^2 j^2\tilde{\mathbf{X}}^{j} = 0.$$

$\mathbf{X}^{o}{}_{\alpha}$ and $\mathbf{X}^{o}$ do not depend on $\phi$. The notation $z^{\alpha}$ stands for

$$z_1^{\alpha_1}z_2^{\alpha_2}, \alpha_1,\alpha_2 \in \mathbf{N}.$$

It follows from the usual averaging theory (see again (Cho77a) for details) that the solutions of these equations can be approximated by the solutions of the averaged system:

$$\dot{\phi} = 1+\delta\tilde{X}^{o}(r)+\epsilon X^{o}_{o}(r),$$

$$\dot{r} = \delta\tilde{Y}^{o}(r)+\epsilon Y^{o}_{o}(r),$$

$$\dot{z} = \Lambda_2 z.$$

These approximations have $O(\epsilon+\delta)$ -accuracy on the time-interval

$$0\leqslant(\epsilon+\delta)t\leqslant L$$

(for the z-component the interval is $[0,\infty)$).

Clearly, this estimate is sharpest if $\epsilon$ and $\delta$ are of the same order of magnitude.

If the averaged equation has an attracting limit-cycle as a solution, then in the domain of attraction the time-interval of validity of the $O(\epsilon+\delta)$ -estimate is $[0,\infty)$ for the r and z component.

This makes it possible, in principle, to obtain estimates for the $\phi$ -component on arbitrary long time-scales (in powers of $\epsilon$, that is) by simply computing higher order averaged vectorfields.

We shall not follow this line of thought here, due to the considerable amount of work and the fact that the results can never be spectacular, since it can only be a regular perturbation of the approximations which we are going to find (This follows from the first order averaged equations and represents the generic case; in practice one may meet exceptions).

After some calculations, we find the following averaged equations for our problem:

$$\dot{\phi} = 1+\frac{41\delta}{2\Delta 7^2} + \frac{3\epsilon r^2}{8\Delta 7^3}\left( - \frac{41\cdot 109}{8\cdot 7^2} + \frac{517B^2}{4\cdot 7} + 7B^4\right),$$

$$\dot{r} = \frac{2B\delta r}{7\Delta} - \frac{\epsilon B r^3}{8\Delta 7^3}\left(\frac{10441}{4\cdot 7^2} + \frac{277B^2}{14}\right),$$

$$\dot{z} = \Lambda_2 z.$$

It is, of course, easy to solve this equation directly, but it is more fun to obtain an asymptotic approximation for large t, without actually solving it:

Consider , with new coefficients $\alpha,\beta,\gamma,\delta \in \mathbf{R}$

$$\dot{\phi} = 1+\gamma+\delta r^2, \qquad \phi(0) = \phi_o,$$

$$\dot{r} = \alpha r - \beta r^3 , \qquad r(0) = r_o.$$

Let

$$r_\infty^2 = \frac{\alpha}{\beta},$$

then

$$\frac{d}{d\tau}\log r = \beta(r_\infty^2 - r^2)$$

and

$$\log\frac{r_\infty}{r_o} = \beta\int_0^\infty (r_\infty^2 - r^2(t))dt.$$

Clearly

$$\dot{\phi} = 1+\gamma+\delta r^2 = 1+\gamma+\delta r_\infty^2 + \delta(r^2 - r_\infty^2)$$

or

$$\phi(t) = \phi_o + (1+\gamma+\delta r_\infty^2)t + \delta\int_0^t (r^2(t) - r_\infty^2)dt$$

$$= \phi_o + (1+\gamma+\delta r_\infty^2)t + \delta\int_0^\infty (r^2(t) - r_\infty^2)dt - \delta\int_t^\infty (r^2(t) - r_\infty^2)dt$$

$$= \phi_o + (1+\gamma+\delta r_\infty^2)t + \frac{\delta}{\beta}\log\frac{r_o}{r_\infty} - \delta\int_t^\infty (r^2(t) - r_\infty^2)dt.$$

Now

$$r^2(t) - r_\infty^2 = O(e^{-2\alpha t}) \text{ for } t\to\infty,$$

which is clear from the equation for $r^2$ and the Lyapunov-stability estimate, so

$$\phi(t) = \phi_o + \frac{\delta}{\beta}\log\frac{r_o}{r_\infty} + (1+\gamma+\delta r_\infty^2)t + O(e^{-2\alpha t}).$$

The phase-shift $\frac{\delta}{\beta}\log\frac{r_o}{r_\infty}$ and especially the frequency-shift $\gamma+\delta r_\infty^2$, can be used to check the asymptotic computational results numerically, and to check the numerical results, by extrapolation in $\epsilon$, asymptotically.

**References**

Bel70a. Bellman,R., *Methods of Nonlinear Analysis, I,* Academic Press, New York (1970).

Cho77a. Chow,S.-N. and Mallet-Paret,J., "Integral averaging and bifurcation," *Journal of Differential Equations* **26**, pp. 112-159 (1977).

Set77a. Sethna,P.R. and Schapiro,S.M., "Nonlinear behaviour of flutter unstable dynamical systems with gyroscopic and circulatory forces," *J. Applied Mechanics* **44**, pp. 755-762 (1977).

# 8.7. Some Elementary Exercises in Celestial Mechanics

## 8.7.1. Introduction

For centuries celestial mechanics has been an exceptional rich source of problems and results in mathematics. To some extent this is still the case. Today one can discern, rather artificially, three problem fields. The first one is the study of classical problems like perturbed *Kepler* motion, orbits in the three-body problem, the theory of asteroids and comets, etc. The second one is a small but relatively important field in which the astrophysicists are interested; we are referring to systems with evolution like for instance changes caused by tidal effects or by exchange of mass. The third field is what one could call 'mathematical celestial mechanics', a subject which is part of the theory of dynamical systems. The distinction between the fields is artificial. There is some interplay between the fields and hopefully, this will increase in the future. An interesting example of a study combining the first and the third field is the paper by *Brjuno* (Brj70a). A typical example of an important mathematical paper which should still find its use in classical celestial mechanics is *Moser*'s study on the geometrical interpretation of the *Kepler* problem (Mos70a). Surveys of mathematical aspects of celestial mechanics have been given in (Mos73a) and (Ale81a).

Here we shall be concerned with simple examples of the use of averaging theory. Apart from being an exercise it may serve as an introduction to the more complicated general literature. One of the difficulties of the literature is the use of many different coordinate systems. We have chosen here for the perturbed harmonic oscillator formulation which eases averaging and admits a simple geometric interpretation. For reasons of comparison we shall demonstrate the use of another coordinate system in a particular problem.

In celestial mechanics thousands of papers have been published and a large number of elementary results are being rediscovered again and again. Our reference list will therefore do no justice to all scientists whose efforts were directed towards the problems mentioned here. For a survey of theory and results see (Stu74a) and (Hag76a). However, also there the reference lists are far from complete and the mathematical discussion is sometimes confusing.

### 8.7.2. The unperturbed Kepler problem

Consider the motion of two point masses acting upon each other by *Newtonian* force fields. In relative coordinates $\mathbf{r}$, with norm $r=\|\mathbf{r}\|$ we have for the gravitational potential

$$V_o(\mathbf{r})=-\frac{\mu}{r}, \qquad \text{A7-1}$$

with $\mu$ the gravitational constant; the equations of motion are

$$\ddot{\mathbf{r}}=-\frac{\mu}{r^3}\mathbf{r}. \qquad \text{A7-2}$$

The angular momentum vector

$$\mathbf{h}=\mathbf{r}\times\dot{\mathbf{r}} \qquad \text{A7-3}$$

is an (vectorvalued) integral of motion; this follows from

$$\frac{d\mathbf{h}}{dt}=\dot{\mathbf{r}}\times\dot{\mathbf{r}}+\mathbf{r}\times\ddot{\mathbf{r}}=\mathbf{r}\times(-\frac{\mu}{r^3}\mathbf{r})=0.$$

The energy of the system

$$E=½\|\dot{\mathbf{r}}\|^2+V_o(r) \qquad \text{A7-4}$$

is also an integral of motion. Note that the equations of motion represent a three degree of freedom system, derived from a *Hamiltonian*. Three independent integrals suffice to make the system integrable. The integrals (A7-3) and (A7-4) however represent already four independent integrals which implies that the integrability of the unperturbed *Kepler* problem is characterized by an unusual degeneration. We recognize this also by concluding from the constancy of the angular momentum vector $\mathbf{h}$ that the orbits are planar. Choosing for instance $z(0)=\dot{z}(0)=0$ implies that $z(t),\dot{z}(t)$ are zero for all time. It is then natural to choose such initial conditions and to introduce polar coordinates $x=r\cos(\phi)$, $y=r\sin(\phi)$ in the plane. Equation (A7-2) yields

$$\ddot{r}-r\dot{\phi}^2=-\frac{\mu}{r^2}, \qquad \text{A7-5}$$

$$\frac{d}{dt}(r^2\dot{\phi})=0. \qquad \text{A7-6}$$

The last equation corresponds with the component of the angular momentum vector which is unequal to zero and we have

$$r^2\dot{\phi}=h \qquad \text{A7-7}$$

with $h=\|\mathbf{h}\|$. We could solve equation (A7-5) in various ways which have all some special advantages. If $E<0$, the orbits are periodic and they describe conic sections. A direct description of the orbits is possible by using geometric variables like the *eccentricity* $e$, *semi-major axis* $a$ and dynamical variables like the period $P$, the time of *peri-astron passage* $T$, etc.

Keeping an eye on perturbation theory a useful presentation of the solution is the harmonic oscillator formulation. Introduce $\phi$ as a time-like variable and put

$$u=\frac{1}{r}. \qquad \text{A7-8}$$

Transforming $r,t\mapsto u,\phi$ in equation (A7-5) produces

$$\frac{d^2u}{d\phi^2}+u=\frac{\mu}{h^2}. \qquad \text{A7-9}$$

The solution can be written as

$$u=\frac{\mu}{h^2}+\alpha\cos(\phi+\beta) \qquad \text{A7-10a}$$

or equivalently

$$u=\frac{\mu}{h^2}+A\cos(\phi)+B\sin(\phi), \qquad \text{A7-10b}$$

with $\alpha,\beta,A,B\in\mathbf{R}$.

### 8.7.3. Perturbations

In the sequel we shall consider various perturbations of the *Kepler* problem. One of those is to admit variation of $\mu$ by changes of the gravitational field with time or change of the total mass. These problems will be formulated later on. For an examination of various perturbing forces, see (Gey71a).

In general we can write the equation of the perturbed *Kepler* problem

$$\ddot{\mathbf{r}}=-\frac{\mu}{r^3}\mathbf{r}+\mathbf{F} \qquad \text{A7-11}$$

in which $\mathbf{r}$ is again the relative position vector, $\mu$ the gravitational constant; $\mathbf{F}$ stands for the as yet unspecified perturbation. The angular momentum vector (A7-3) will in general only be constant if $\mathbf{F}$ lies along $\mathbf{r}$ as we have

$$\frac{d\mathbf{h}}{dt}=\mathbf{r}\times\mathbf{F}. \qquad \text{A7-12}$$

It will be useful to introduce spherical coordinates $x=r\cos(\phi)\sin(\theta)$, $y=r\sin(\phi)\sin(\theta)$ and $z=r\cos(\theta)$ in which $\theta$ is the *colatitude*, $\phi$ the *azimuthal angle* in the equatorial plane. Specifying $\mathbf{F}=(F_x,F_y,F_z)$ we find from equation (A7-11)

$$\ddot{r}-r\dot{\phi}^2\sin^2\theta-r\dot{\theta}^2= \qquad \text{A7-13}$$

$$-\frac{\mu}{r^2}+(F_x\cos(\phi)+F_y\sin(\phi))\sin(\theta)+F_z\cos(\theta).$$

It is useful to write this equation in a different form using the angular momentum. From (A7-3) we calculate

$$h^2=\|\mathbf{h}\|^2=r^4\dot{\theta}^2+r^4\dot{\phi}^2\sin^2(\theta).$$

Equation (A7-13) can then be written as

$$\ddot{r}-\frac{h^2}{r^3}=-\frac{\mu}{r^2}+(F_x\cos(\phi)+F_y\sin(\phi))\sin(\theta)+F_z\cos(\theta). \qquad \text{A7-13a}$$

Equation (A7-11) is of order 6 so we need two more second order equations. A combination of the first two components of angular momentum produces

$$\frac{d}{dt}(r^2\dot{\theta})-r^2(\dot{\phi})^2\sin(\theta)\cos(\theta)= \qquad \text{A7-14}$$

$$-rF_z\sin(\theta)+r\cos(\theta)(F_y\sin(\phi)+F_x\cos(\phi)).$$

The third ($z$) component of angular momentum is described by

$$\frac{d}{dt}(r^2\dot{\phi}\sin^2\theta)=-r\sin(\theta)(F_x\sin(\phi)-F_y\cos(\phi)). \qquad \text{A7-15}$$

Note that the components of **F** still have to be rewritten in spherical coordinates.

## 8.7.4. Motion around an 'oblate planet'

To illustrate this formulation of the perturbed *Kepler* problem we consider the case that one of the bodies can be considered a point mass, the other body is axi-symmetric and flattened at the poles. The description of the motion of a point mass around such an *oblate planet* has some relevance for satellite mechanics. Suppose that the polar axis is taken as the $z$-axis and the $x$ and $y$ axes are taken in the equatorial plane. The gravitational potential can be represented by a convergent series

$$V=-\frac{\mu}{r}[1-\sum_{n=2}^{\infty}\frac{1}{r^n}J_nP_n(\frac{z}{r})] \qquad \text{A7-16}$$

where the units are such that the equatorial radius corresponds with $r=1$. The $P_n$ are the standard *Legendre* polynomials of degree $n$, the $J_n$ are constants determined by the axisymmetric distribution of mass (they have nothing to do with *Bessel* functions).

In the case of the planet *Earth* we have

$$J_2=1.1\times10^{-3},$$

$$J_3 = -2.3 \times 10^{-6},$$

$$J_4 = -1.7 \times 10^{-6}.$$

The constants $J_5$, $J_6$ etc. do not exceed the order of magnitude $10^{-6}$. A first order study of satellite orbits around the earth involves the truncation of the series after $J_2$; we put

$$V_1 = -\frac{\mu}{r} + \frac{\mu}{r^3} J_2 P_2\left(\frac{z}{r}\right) = -\frac{\mu}{r} + \tfrac{1}{2} J_2 \frac{\mu}{r^3}\left(1 - 3\frac{z^2}{r^2}\right)$$

or

$$V_1 = -\frac{\mu}{r} + \epsilon \frac{\mu}{r^3}(1 - 3\cos^2(\theta)), \qquad \text{A7-17}$$

where $\epsilon = \tfrac{1}{2} J_2$. Taking the gradient of $V_1$ we find for the components of the perturbation vector

$$F_x = \epsilon \frac{3\mu}{r^5} x\left(-1 + 5\frac{z^2}{r^2}\right),$$

$$F_y = \epsilon \frac{3\mu}{r^5} y\left(-1 + 5\frac{z^2}{r^2}\right),$$

$$F_z = \epsilon \frac{3\mu}{r^5} z\left(-3 + 5\frac{z^2}{r^2}\right).$$

The equations of motion (A7-(13-15)) become in this case

$$\ddot{r} - \frac{h^2}{r^3} = -\frac{\mu}{r^2} - \epsilon \frac{3\mu}{r^4}(1 - 3\cos^2(\theta)), \qquad \text{A7-18}$$

$$\frac{d}{dt}(r^2 \dot{\theta}) - r^2 (\dot{\phi})^2 \sin(\theta)\cos(\theta) = \epsilon \frac{6\mu}{r^3} \sin(\theta)\cos(\theta), \qquad \text{A7-19}$$

$$\frac{d}{dt}(r^2 \dot{\phi} \sin^2(\theta)) = 0. \qquad \text{A7-20}$$

The last equation can be integrated and then expresses that the $z$-component of angular momentum is conserved. This could be expected from the assumption of axial symmetry.

Note that the energy is conserved, so we have two integrals of motion of system (A7-(18-20)). For the system to be integrable we need another independent integral; we have no results available on the existence of such a third integral.

## 8.7.5. Harmonic oscillator formulation for motion around an 'oblate planet'

We shall transform equations (A7-(18-19)) in the following way. The dependent variables $r$ and $\theta$ are replaced by

$$u = \frac{1}{r} \quad \textit{and} \quad v = \cos(\theta). \qquad \text{A7-21}$$

The independent variable $t$ is replaced by a time-like variable $\tau$ given by

$$\dot{\tau}=\frac{h}{r^2}=hu^2 \quad , \ \tau(0)=0; \qquad \text{A7-22}$$

$h$ is again the length of the angular momentum vector. Note that $\tau$ is monotonically increasing, as a time-like variable should, except in the case of radial (vertical) motion. We cannot expect that for all types of perturbations $\tau$ runs ad infinitum like $t$; in other words, equation (A7-22) may define a mapping from $[0,\infty)$ into $[0,C]$ with $C$ a positive constant. If no perturbations are present, $\tau$ represents an angular variable, see 8.5.2.

In the case of equations (A7-(18-19)) we find, using the transformations (A7-(21-22))

$$\frac{d^2u}{d\tau^2}+u=\frac{\mu}{h^2}+\epsilon\frac{6\mu}{h^2}uv\frac{du}{d\tau}\frac{dv}{d\tau}+\epsilon 3\mu\frac{u^2}{h^2}(1-3v^2), \qquad \text{A7-23}$$

$$\frac{d^2v}{d\tau^2}+v=\epsilon\frac{6\mu}{h^2}uv(\frac{dv}{d\tau})^2-\epsilon 6\frac{\mu}{h^2}uv(1-v^2), \qquad \text{A7-24}$$

$$\frac{dh}{d\tau}=-\epsilon\frac{6\mu}{h}uv\frac{dv}{d\tau}. \qquad \text{A7-25}$$

Instead of the variable $h$ it makes sense to use as variable $h^2$ or $\frac{\mu}{h^2}$; here we shall use $h$.

In a slightly different form these equations have been presented by *Kyner* (Kyn65a); note however that the discussion in that paper on the time-scale of validity and the order of approximation is respectively wrong and unnecessarily complicated.

System (A7-(23-25)) still admits the energy integral but no other integrals are available (Exercise: in what sense are our transformations canonical?). Having solved system (A7-(23-25)) we can transform back to time $t$ by solving equation (A7-22).

## 8.7.6. First order averaging for motion around an 'oblate planet'

We have to put equations (A7-(230-25)) in the standard form for averaging using the familiar *Lagrange* method of variation of parameters. As we have seen in chapter 2 the choice of the perturbation formulation affects the computational work, not the final result. we find it convenient to choose in this problem the transformation $u,\frac{du}{d\tau}\mapsto a_1,b_1$ and $v,\frac{dv}{d\tau}\mapsto a_2,b_2$ defined by

$$u=\frac{\mu}{h^2}+a_1\cos(\tau+b_1)\ , \ \frac{du}{d\tau}=-a_1\sin(\tau+b_1), \qquad \text{A7-26}$$

$$v=a_2\cos(\tau+b_2) \quad , \ \frac{dv}{d\tau}=-a_2\sin(\tau+b_2). \qquad \text{A7-27}$$

The inclination $i$ of the orbital plane, which is a constant of motion if no perturbations are present, is connected with the new variable $a_2$ by

$a_2 = \sin(i)$. Abbreviating equations (A7-(23-24)) by

$$\frac{d^2u}{d\tau^2} + u = \frac{\mu}{h^2} + \epsilon G_1,$$

$$\frac{d^2v}{d\tau^2} + v = \epsilon G_2,$$

we find the system

$$\frac{da_1}{d\tau} = \frac{2\mu}{h^3}\frac{dh}{d\tau}\cos(\tau + b_1) - \epsilon G_1 \sin(\tau + b_1), \qquad \text{A7-28}$$

$$\frac{db_1}{d\tau} = -\frac{2\mu}{h^3}\frac{dh}{d\tau}\frac{\sin(\tau + b_1)}{a_1} - \epsilon G_1 \frac{\cos(\tau + b_1)}{a_1},$$

$$\frac{da_2}{d\tau} = -\epsilon G_2 \sin(\tau + b_2), \qquad \text{A7-29}$$

$$\frac{db_2}{d\tau} = -\epsilon G_2 \frac{\cos(\tau + b_2)}{a_2}.$$

We have to add equation (A7-25) and in $G_1$, $G_2$ we have to substitute variables according to equations (A7-(26-27)). The resulting system is $2\pi$-periodic in $\tau$ and we apply first order averaging. The approximations of $a$, $b$ will be indicated by $\alpha$, $\beta$. For the right hand side of equation (A7-25) we find average zero, so that $h(\tau) = h(0) + O(\epsilon)$ on the time-scale $\frac{1}{\epsilon}$, i.e. for $0 \leqslant \epsilon\tau \leqslant L$ with $L$ a constant independent of $\epsilon$. Averaging of equations (A7-(28-29)) produces

$$\frac{d\alpha_1}{d\tau} = 0, \qquad \text{A7-30}$$

$$\frac{d\alpha_2}{d\tau} = 0,$$

$$\frac{d\beta_1}{d\tau} = -\epsilon \frac{3\mu^2}{h^4(0)}(1 - \frac{3}{2}\alpha_2^2),$$

$$\frac{d\beta_2}{d\tau} = \epsilon \frac{3\mu^2}{h^4(0)}(1 - \alpha_2^2).$$

So in this approximation the system is integrable. Putting $p = -\frac{3\mu^2}{h^4(0)}(1 - \frac{3}{2}\alpha_2^2(0))$ and $q = \frac{3\mu^2}{h^4(0)}(1 - \alpha_2^2(0))$ we conclude that we have obtained the following first order approximations on the time-scale $\frac{1}{\epsilon}$

$$u(\tau) = \frac{\mu}{h^2(0)} + a_1(0)\cos(\tau + \epsilon p\tau + b_1(0)) + O(\epsilon), \qquad \text{A7-31}$$

$$v(\tau) = a_2(0)\cos(\tau + \epsilon q\tau + b_2(0)) + O(\epsilon).$$

It is remarkable that the original, rather complex, perturbation problem admits such simple approximations. Note however that in higher approximation or on longer time-scales qualitatively new phenomena may occur. To illustrate this we shall discuss some special solutions.

**Equatorial orbits**

The choice of potential (A7-16) has as a consequence that we can restrict the motion to the equator plane $z=0$, or $\theta=\frac{1}{2}\pi$ for all time. In equation (A7-24) these solutions correspond with $v=\frac{dv}{d\tau}=0,\ \tau\geqslant 0$. Equation (A7-23) reduces to

$$\frac{d^2u}{d\tau^2}+u=\frac{\mu}{h^2}+\epsilon\frac{3\mu}{h^2}u^2. \qquad \text{A7-32}$$

The time-like variable $\tau$ can be identified with the *azimuthal angle* $\phi$. Equation (A7-25) produces that $h$ is a constant of motion in this case. It is not difficult to show that the solutions of the equations of motion restricted to the equatorial plane are periodic. We find

$$u(\tau)=\frac{\mu}{h^2}+a_1(0)\cos(\tau-\epsilon\frac{3\mu^2}{h^4}\tau+b_1(0))+O(\epsilon)$$

on the time-scale $\frac{1}{\epsilon}$. Using the theory of chapter 3 it is very easy to obtain higher order approximations but no new qualitative phenomena can be expected at higher order.

**Polar orbits**

The axi-symmetry of the potential (A7-16) triggers off the existence of orbits in meridional planes: taking $\dot{\phi}=0$ for all time solves equation (A7-20) (and more in general (A7-15) with the assumption of axi-symmetry). In this case the time-like variable $\tau$ can be identified with $\theta$; equation (A7-24) is solved by $v(\tau)=\cos(\tau)$. System (A7-31) produces the approximation $(s=\frac{3\mu^2}{2h^4(0)})$

$$u(\tau)=\frac{\mu}{h^2(0)}+a_1(0)\cos(\tau+\epsilon s\tau+b_1(0))+O(\epsilon)$$

on the time-scale $\frac{1}{\epsilon}$. Again one can obtain higher order approximations for the third order system (A7-(23-25)).

**The critical inclination problem**

Analyzing the averaged system (A7-30) one expects a resonance domain near the zeros of $\frac{d\beta_1}{d\tau}-\frac{d\beta_2}{d\tau}$. This expectation is founded on our analysis of averaging over spatial variables (Chapter 5) and the theory of higher order resonance in two degrees of freedom *Hamiltonian* systems (Chapter 7). The resonance domain follows from

$$\frac{d\beta_1}{d\tau}-\frac{d\beta_2}{d\tau}=\epsilon\frac{3\mu^2}{h^4(0)}(\frac{5}{2}\alpha_2^2-2)=0$$

or $\alpha_2^2 = \frac{4}{5}$; in terms of the inclination i

$$\sin^2(i) = \frac{4}{5}.$$

This $i$ is called the *critical inclination.* To analyze the flow in the resonance domain we have to use higher order approximations. Using different transformations, but related techniques the higher order problem has been discussed by various authors; we mention (Eck66a) and (Dep81a).

## 8.7.7. A dissipative force: atmospheric drag

In this section we shall study the influence of a dissipative force by introducing atmospheric drag. In the subsequent sections we introduce other dissipative forces producing evolution of two body systems.

In the preceding sections we studied *Hamiltonian* perturbations of an integrable *Hamiltonian* system. The introduction of dissipative forces presents qualitatively new phenomena; an interesting aspect is however that we can apply the same perturbation techniques.

Suppose that the second body is moving through an atmosphere surrounding the primary body. We ignore the lift acceleration or assume that this effect is averaged out by tumbling effects. For the drag acceleration vector we assume that it takes the form

$$-\epsilon B(r)|\dot{\mathbf{r}}|^m \dot{\mathbf{r}}, \qquad m \ \text{a constant.}$$

$B(r)$ is a positive function, determined by the density of the atmosphere; in more realistic models $B$ also depends on the angles and the time. Often one chooses $m=1$, corresponding with a velocity-squared aerodynamic force law; $m=0$, $B$ constant, corresponds with linear friction. (Some aspects of aerodynamic acceleration, including lift effects, and the corresponding harmonic oscillator formulation have been discussed in (Woe72a); for perturbation effects in atmospheres see also (Gey71a)).

Assuming that a purely gravitational perturbation force $\mathbf{F}_g$ is present, we have in equation (A7-11)

$$\mathbf{F} = \epsilon \mathbf{F}_g - \epsilon B(r)|\dot{\mathbf{r}}|^m \dot{\mathbf{r}}.$$

To illustrate the treatment we restrict ourselves to equatorial orbits. This only means a restriction on $\mathbf{F}_g$ to admit the existence of such orbits (as in the case of motion around an oblate planet). Putting $z=0$ ($\theta = \frac{1}{2}\pi$) we find with $h = r^2\dot{\phi}$, $|\dot{\mathbf{r}}|^2 = \dot{r}^2 + h^2 r^{-2}$ from equation (A7-13a) and (A7-15)

$$\ddot{r} - \frac{h^2}{r^3} = -\frac{\mu}{r^2} + \epsilon f_1(r,\phi) - \epsilon B(r)(\dot{r}^2 + h^2 r^{-2})^{\frac{m}{2}} \dot{r}, \qquad \text{A7-33}$$

$$\frac{dh}{dt} = \epsilon f_2(r\phi) - \epsilon B(r)(\dot{r}^2 + h^2 r^{-2})^{\frac{m}{2}} h, \qquad \text{A7-34}$$

in which $f_1$ and $f_2$ are gravitational perturbations to be computed from

$$f_1(r,\phi)=F_{gx}\cos(\phi)+F_{gy}\sin(\phi),$$

$$f_2(r,\phi)=-r(F_{gx}\sin(\phi)-f_{gy}\cos(\phi)).$$

The requirement of continuity yields that $f_1$ and $f_2$ are $2\pi$-periodic in $\phi$. For example in the case of motion around an oblate planet we have, comparing with equations (A7-18) and (A7-19), $f_1(r,\phi)=-\frac{3\mu}{r^4}$, $f_2(r,\phi)=0$. The time-like variable $\tau$, introduced by equation (A7-22), can be identified with $\phi$; putting $u=\frac{1}{r}$ we find

$$\frac{d^2u}{d\tau^2}+u=\frac{\mu}{h^2}-\epsilon\frac{f_1(\frac{1}{u},\tau)}{h^2u^2}-\epsilon\frac{f_2(\frac{1}{u},\tau)}{h^2u^2}\frac{du}{d\tau}, \qquad \text{A7-35}$$

$$\frac{dh}{d\tau}=-\epsilon\frac{f_2(\frac{1}{u},\tau)}{hu^2}-\epsilon\frac{B(\frac{1}{u})}{u^2}[u^2+(\frac{du}{d\tau})^2]^{\frac{m}{2}}h^m. \qquad \text{A7-36}$$

The perturbation problem (A7-(35-36)) can be treated by first order averaging as in 8.6. After specifying $f_1$, $f_2$, $B$ and $m$ we transform by (A7-26)

$$u=\frac{\mu}{h^2}+a_1\cos(\tau+b_1)\,,\quad \frac{du}{d\tau}=-a_1\sin(\tau+b_1).$$

To be more explicit we discuss the case of equatorial motion around an oblate planet; equations (A7-(35-36)) become

$$\frac{d^2u}{d\tau^2}+u=\frac{\mu}{h^2}+3\epsilon\mu\frac{u^2}{h^2}, \qquad \text{A7-37}$$

$$\frac{dh}{d\tau}=-\epsilon\frac{B(\frac{1}{u})}{u^2}[u^2+(\frac{du}{d\tau})^2]^{\frac{m}{2}}h^m. \qquad \text{A7-38}$$

As the density function $B$ is positive, it is clear that the length of the angular momentum vector $h$ is monotonically decreasing. Transforming we have (Cf. equation (A7-28))

$$\frac{da_1}{d\tau}=-\epsilon\frac{2\mu}{u^2}B(\frac{1}{u})h^{m-3}[\frac{\mu^2}{h^4}+a_1^2+2\frac{\mu}{h^2}a_1\cos(\tau+b_1)]^{\frac{m}{2}}\cos(\tau+b_1)$$

$$-3\epsilon\mu\frac{u^2}{h^2}\sin(\tau+b_1), \qquad \text{A7-39}$$

$$\frac{db_1}{d\tau}=+\epsilon\frac{2\mu}{u^2}B(\frac{1}{u})h^{m-3}[\frac{\mu^2}{h^4}+a_1^2+2\frac{\mu}{h^2}a_1\cos(\tau+b_1)]^{\frac{m}{2}}\frac{\sin(\tau+b_1)}{a_1}$$

$$-3\epsilon\mu\frac{u^2}{h^2}\frac{\cos(\tau+b_1)}{a_1} \qquad \text{A7-40}$$

to which we add equation (A7-38); at some places we still have to write down the expression for $u$. The right hand side of equations (A7-(38-40)) is $2\pi$-periodic in $\tau$; averaging produces that the second term on the right hand side of (A7-39) and the first term on the right hand side of (A7-40) vanishes. For the density function $B$ one usually chooses a function exponentially decreasing with $r$ or a combination of powers of $r$ (hyperbolic density law). A simple case arises if we choose

$$B(r)=\frac{B_o}{r^2}.$$

The averaged equations take the form

$$\frac{d\alpha_1}{d\tau}=-2\epsilon\mu B_o\mathbf{h}^{m-3}\frac{1}{2\pi}\int_0^{2\pi}[\frac{\mu^2}{\mathbf{h}^4}+\alpha_1^2+2\frac{\mu}{\mathbf{h}^2}\alpha_1\cos(\tau+\beta_1)]^{\frac{m}{2}}\cos(\tau+\beta_1)d\tau,$$

$$\frac{d\beta_1}{d\tau}=-3\epsilon\frac{\mu^2}{\mathbf{h}^4}, \qquad \text{A7-41}$$

$$\frac{d\mathbf{h}}{d\tau}=-\epsilon B_o\mathbf{h}^m\frac{1}{2\pi}\int_0^{2\pi}[\frac{\mu^2}{\mathbf{h}^4}+\alpha_1^2+2\frac{\mu}{\mathbf{h}^2}\alpha_1\cos(\tau+\beta_1)]^{\frac{m}{2}}d\tau,$$

in which $\alpha_1$, $\beta_1$, $\mathbf{h}$ are $O(\epsilon)$ approximations of $a_1$, $b_1$ and $h$ on the time-scale $\frac{1}{\epsilon}$, assuming that we impose the initial conditions $\alpha_1(0)=a_1(0)$ etc. In the case of a velocity-squared aerodynamic force law ($m=1$), we have still to evaluate two definite integrals. These integrals are elliptic and they can be analyzed by series expansion (note that $0<2\frac{\mu}{\mathbf{h}^2}\alpha_1<\frac{\mu^2}{\mathbf{h}^4}+\alpha_1^2$ so that we can use binomial expansion).

Linear force laws are of less practical interest but it can still be instructive to carry out the calculations. If $m=0$ we find

$$\alpha_1(\tau)=a_1(0)\ ,\ \beta_1(\tau)=b_1(0)+\tfrac{3}{4}\frac{\mu^2}{B_o}(e^{-4\epsilon B_o\tau}-1)\ ,\ \mathbf{h}=h(0)e^{-\epsilon B_o\tau}$$

which in the original variables corresponds with a spiraling down of the body moving through atmosphere.

## 8.7.8. Systems with mass loss or variable $G$

We consider now a class of problems in which mass is ejected isotropically from the two-body system and is lost to the system or in which the gravitational 'constant' $G$ decreases with time. The treatment is taken from (Ver75a).

It can be shown that the relative motion of the two bodies takes place in a plane; introducing polar coordinates in this plane we have the equations of motion

$$\ddot{r}=-\frac{\mu(t)}{r^2}+\frac{h^2}{r^3}, \qquad \text{A7-42}$$

$$r^2\dot{\phi}=h. \qquad \text{A7-43}$$

The length of the angular momentum vector is conserved, $\mu(t)$ is a monotonically decreasing function with time.

Introducing again $u=\frac{1}{r}$ and the time-like variable $\tau$ by $\dot{\tau}=hu^2$ we find

$$\frac{d^2u}{d\tau^2}+u=\frac{\mu(t(\tau))}{h^2}. \qquad \text{A7-44}$$

We assume that $\mu$ varies slowly with time; in particular we shall assume that

$$\dot{\mu}=-\epsilon\mu^3 \ , \ \mu(0)=\mu_o. \qquad \text{A7-45}$$

In the reference given above one can find a treatment of a more general class of functions $\mu$, and also the case of fast changes in $\mu$ has been discussed there. We can write the problem (A7-(44-45)) as

$$\frac{d^2u}{d\tau^2}+u=w, \qquad \text{A7-46}$$

$$\frac{dw}{d\tau}=-\epsilon h^3\frac{w^3}{u^2}, \qquad \text{A7-47}$$

where we put $\frac{\mu(t(\tau))}{h^2}=w(\tau)$. To obtain the standard form for averaging it is convenient to transform $u,\frac{du}{d\tau}\mapsto a,b$ by

$$u=w+a\cos(\tau)+b\sin(\tau) \ , \ \frac{du}{d\tau}=-a\sin(\tau)+b\cos(\tau). \qquad \text{A7-48}$$

We find

$$\frac{da}{d\tau}=\epsilon h^3\frac{w^3\cos(\tau)}{(w+a\cos(\tau)+b\sin(\tau))^2} \ , \ a(0) \ given,$$

$$\frac{db}{d\tau}=\epsilon h^3\frac{w^3\sin(\tau)}{(w+a\cos(\tau)+b\sin(\tau))^2} \ , \ b(0) \ given, \qquad \text{A7-49}$$

$$\frac{dw}{d\tau}=-\epsilon h^3\frac{w^3}{(w+a\cos(\tau)+b\sin(\tau))^2} \ , \ w(0)=\frac{\mu(0)}{h^2}.$$

The right hand side of system (A7-49) is $2\pi$-periodic in $\tau$ and averaging produces

$$\frac{d\alpha}{d\tau}=-\epsilon h^3\frac{\alpha W^3}{(W^2-\alpha^2-\beta^2)^{\frac{3}{2}}} \ , \ \alpha(0)=a(0),$$

$$\frac{d\beta}{d\tau}=-\epsilon h^3\frac{\beta W^3}{(W^2-\alpha^2-\beta^2)^{\frac{3}{2}}} \ , \ \beta(0)=b(0), \qquad \text{A7-50}$$

$$\frac{dW}{d\tau}=-\epsilon h^3\frac{W^4}{(W^2-\alpha^2-\beta^2)^{\frac{3}{2}}} \ , \ W(0)=w(0),$$

where $a(\tau)-\alpha(\tau)$, $b(\tau)-\beta(\tau)$, $w(\tau)-W(\tau)=O(\epsilon)$ on the time-scale $\frac{1}{\epsilon}$. It is easy to see that

$$\frac{\alpha(\tau)}{\alpha(0)} = \frac{\beta(\tau)}{\beta(0)} = \frac{W(\tau)}{w(0)}, \quad (a(0), b(0) \neq 0) \qquad \text{A7-51}$$

and we find

$$\alpha(\tau) = a(0)e^{-\epsilon\lambda\tau},$$

with $\lambda = \dfrac{h^3 w^3(0) a^2(0)}{(w^2(0) - a^2(0) - b^2(0))^{\frac{3}{2}}}$. Using again (A7-51) we can construct an $O(\epsilon)$-approximation for $u(\tau)$ on the time-scale $\dfrac{1}{\epsilon}$. Another possibility is to realize that $W(\tau) = \dfrac{\mu(t(\tau))}{h^2} + O(\epsilon)$ so that with equation ( A7-45)

$$W = \frac{1}{h^2}\left(\frac{1}{\mu(0)^2} + 2\epsilon t\right)^{-\frac{1}{2}} + O(\epsilon) \qquad \text{A7-52}$$

and corresponding expressions for $\alpha$ and $\beta$.

We have performed our calculations without bothering about the conditions of the averaging theorem, apart from periodicity. It follows from the averaged equation (A7-50) that the quantity $(W^2 - \alpha^2 - \beta^2)$ should not be small with respect to $\epsilon$. This condition is not a priori clear from the original equation (A7-49).

Writing down the expression for the instantaneous energy of the two-body system we have

$$E(t) = \tfrac{1}{2}\dot{r}^2 + \tfrac{1}{2}\frac{h^2}{r^2} - \frac{\mu(t)}{r},$$

or, with transformation (A7-48),

$$E(t(\tau)) = \tfrac{1}{2}h^2\left(\left(\frac{du}{d\tau}\right)^2 + u^2\right) - \mu u = \tfrac{1}{2}h^2(a^2 + b^2 - w^2). \qquad \text{A7-53}$$

Negative values of the energy correspond with bound orbits, zero energy with a parabolic orbit, positive energy means escape. The condition that $W^2 - \alpha^2 - \beta^2$ is not small with respect to $\epsilon$ implies that we have to exclude nearly-parabolic orbits among the initial conditions which we study. This condition is reflected in the conditions of the averaging theorem, see for instance Theorem 2.6.1, condition (d).

Note also that, starting in a nearly-parabolic orbit the approximate solutions never yield a positive value of the energy. The conclusion however, that the process described here cannot produce escape orbits is not justified as the averaging method does not apply to the nearly-parabolic transition between elliptic and hyperbolic orbits.

To analyze the situation of nearly-parabolic orbits it is convenient to introduce another coordinate system involving the orbital elements $e$ (eccentricity) and $E$ (eccentric anomaly) or $f$ (true anomaly). We discussed such a system briefly in Example 5.4.5. A rather intricate asymptotic analysis of the nearly-parabolic case shows that nearly all solutions starting there become hyperbolic on a time-scale of order 1 with respect to $\epsilon$; see (Laa72a).

## 8.7.9. Two-body system with increasing mass

Dynamically this is a process which is very different from the process of decrease of mass. In the case of decrease of mass each layer of ejected material takes with it an amount of instantaneous momentum. In the case of increase of mass, we have to make an assumption about the momentum of the material falling in. With certain assumptions, see (Ver75a), the orbits are found again in a plane. Introducing the total mass of the system $m = m(t)$ and the gravitational constant $G$ we then have the equations of motion

$$\ddot{r} = -\frac{Gm}{r^2} + \frac{h^2}{m^2 r^3} - \frac{\dot{m}}{m}\dot{r}, \qquad \text{A7-54}$$

$$mr^2\dot{\phi} = h. \qquad \text{A7-55}$$

Note that (A7-55) represents an integral of motion with a time-varying factor. Introduction of $u = \frac{1}{r}$ and the time-like variable $\tau$ by

$$\dot{\tau} = \frac{hu^2}{m}, \quad \tau(0) = 0$$

we find that the equation

$$\frac{d^2u}{d\tau^2} + u = w \qquad \text{A7-56}$$

in which the equation for $w$ has to be derived from $w = \frac{Gm^3}{h^2}$. Assuming slow increase of mass according to the relation $\dot{m} = \epsilon m^n$ ( $n$ a constant) we find

$$\frac{dw}{d\tau} = \epsilon \frac{3h^{\frac{2n}{3}-1}}{G^{\frac{n}{3}}} \frac{w^{1+\frac{n}{3}}}{u^2}. \qquad \text{A7-57}$$

It is clear that we can approximate the solutions of system (A7-(56-57)) with the same technique as in the preceding section. The approximations can be computed as functions of $\tau$ and $t$ and they represent $O(\epsilon)$ approximations on the time-scale $\frac{1}{\epsilon}$.

**References**

Ale81a. Alekseev, V.M., "Quasirandom oscillations and qualitative questions in celestial mechanics," *Amer. Math. Soc. Transl.* **116**(2), pp. 97-169 (1981).

Brj70a. Brjuno,A.D., "Instability in a *Hamiltonian* system and the distribution of asteroids," *Mat.Sbornik* **83**, pp. 271-312 (1970).

Dep81a. Deprit,A., "The elimination of the parallax in satellite theory," *Celestial Mechanics* **24**, pp. 111-153 (1981).

Eck66a. Eckstein,M., Shi,Y.Y., and Kevorkian,J., "Satellite motion for arbitrary eccentricity and inclination around the smaller primary in the restricted three-body problem," *The Astronomical Journal* **71**, pp. 248-263 (1966).

Gey71a. Geyling,F.T. and Westerman,H.R., *Introduction to Orbital Mechanics,* Addison-Wesley Publ. Co., Reading, Mass. (1971).

Hag76a. Hagihara,Y., *Celestial Mechanics,* The MIT Press, Cambridge, Mass. (1970-76).

Kyn65a. Kyner,W.T., "A mathematical theory of the orbits about an oblate planet," *SIAM J.* **13**(1), pp. 136-171 (1965).

Laa72a. van der Laan,L. and Verhulst,F., "The transition from elliptic to hyperbolic orbits in the two-body problem by slow loss of mass," *Celestial Mechanics* **6**, pp. 343-351 (1972).

Mos70a. Moser,J., "Regularization of Kepler's problem and the averaging method on a manifold," *Comm. Pure Appl. Math.* **23**, pp. 609-636 (1970).

Mos73a. Moser,J., "Stable and random motions in dynamical systems, with special emphasis on celestial mechanics," *Ann. Math. Studies*, Princeton, N.J. **77**, Princeton Univ. Press (1973).

Stu74a. Stumpff,K., *Himmelsmechanik,* VEB Deutscher Verlag der Wissenschaften, Berlin (1959-74).

Ver75a. Verhulst,F., "Asymptotic expansions in the perturbed two-body problem with application to systems with variable mass," *Celestial Mechanics* **11**, pp. 95-129 (1975).

Woe72a. van Woerkom,P.Th.L.M., "A multiple variable approach to perturbed aerospace vehicle motion," Thesis, Princeton (1972).

# 8.8. Informal Introduction to Concepts from (Global) Analysis

## 8.8.1. Introduction

In this appendix we describe briefly a number of concepts which arise frequently in the theory of dynamical systems. Mathematically speaking, this will only serve to give the reader some vague notion what the meaning of some term is, but this may be all one needs. Good introductions are the books (Br73a) and (Abr83a).

## 8.8.2. Manifolds

A manifold is a topological space which locally looks like $\mathbf{R}^n$. This can be mode somewhat more precise as follows: A manifold M is a (reasonable) topological space together with a collection of maps $\{g_\alpha\}_{\alpha \in A}$ (an atlas)

$$g_\alpha : U_\alpha \rightarrow \mathbf{R}^n \ , \quad U_\alpha \ \textit{open in } M \ , \ \ \alpha \in A$$

such that $g_\alpha$ is a bijection, the open neighborhoods $U_\alpha$ cover $M$ and maps with a common domain of definition (i.e. $U_\alpha \cap U_\beta \neq \emptyset$) agree in the following way: the map $g_\alpha g_\beta^{-1}$ is an isomorphism of $g_\beta(U_\alpha \cap U_\beta) \rightarrow g_\alpha(U_\alpha \cap U_\beta)$. If the isomorphism is $C^p$ we speak of a $C^p$-manifold, where $p$ can be a natural number, $\infty$ or $\omega$, i.e. the isomorphism is analytic. The components of the maps $g_\alpha$ are called *local coordinates.*

Some manifolds have their own special notation like $S^n$: the n-*sphere;* this is a sphere in $\mathbf{R}^{n+1}$ given by

$$\sum_{i=1}^{n+1} x_i^2 = 1$$

which is n-dimensional. For instance $S^1$: this is the circle; the ordinary sphere in $\mathbf{R}^3$ is the 2-sphere $S^2$. Another example of a manifold is the n-*torus* $\mathbf{T}^n = \mathbf{R}^n / \mathbf{Z}^n$; this can also be considered as the n-fold product of $S^1$. Thus $\mathbf{T}^2 = S^1 \times S^1$.

A map $f$ between two manifolds is called differentiable, if the induced map $h_\alpha f g_\alpha^{-1}$ is differentiable, where $g_\alpha$ is defined on $U_\alpha$ and $h_\alpha$ on $F(U_\alpha)$. Two $C^1$-manifolds are diffeomorphic if there exists a $C^1$-map $f$ between them such that $f$ is bijective and $f^{-1}$ is also $C^1$. Such a map is called a *diffeomorphism.*

### 8.8.3. Mappings and vector fields

We are familiar with the idea that a tangent vector to a surface (in $\mathbf{R}^n$) is identified with the velocity vector of a curve in the surface. The set of tangent vectors to a surface is abstracted to the concept of tangent bundle $TM$ of a manifold $M$. It is useful to keep in mind the standard examples of classical geometry. In the case of the n-sphere $S^n \subset \mathbf{R}^{n+1}$ the collection of tangent planes to $S^n$ form the tangent bundle $TS^n$; as a set this consists of the sphere itself, with attached to it at each point $x$ a tangent space $T_x S^n$ of dimension $n$. This set is made into a manifold by specifying the maps which locally identify it with $\mathbf{R}^{2n-1}$. In general, if $g_\alpha$ is such a map (or chart) for the original n-manifold $M$, one defines $Tg_\alpha$ by

$$Tg_\alpha(x,v) = (g_\alpha(x), Dg_\alpha(x)v),$$

with $x \in M$ and $v \in \mathbf{R}^n$. A tangent bundle $TM$ is an example of a vector-bundle. Another example is the *cotangent bundle* $T^\star M$, defined by taking instead of the tangent space the dual space (as a vector space, i.e. the space of linear functionals). The reader is encouraged to compute the local maps in this case. The cotangent bundle is of importance in the *Hamiltonian* formulation of mechanics; this we shall illustrate in the next section of this appendix.

If $E$ and $F$ are vector spaces, we call $E \times F$ the *product space.* For $x \in E$ we call $\{x\} \times F$ the *fiber* over $x$. In the tangent bundle $TM$ we denote the fiber above $x$ by $T_x M$.

The vector fields which we are studying very often depend on special parameters or variables like the vectorfield in the equation

$$\dot{x} = f(t,x,\mu) \quad , \; t \in \mathbf{R} \; , \; x \in \mathbf{R}^n \; , \; \mu \in \mathbf{R}^m.$$

We can 'remove' such parameters and variables by *suspension:* In this example we can put

$$\dot{y} = g(y) \quad , \; y \in \mathbf{R}^{n+2},$$

with $y = (t,x,\mu)$ and

$$\dot{x} = f(t,x,\mu),$$

$$\dot{t}=1,$$

$$\dot{\mu}=0.$$

Sometimes we need less trivial suspensions.

Suppose now that we have two manifolds, $M$ and $N$, and a mapping $\phi:M\to N$. Suppose also that we have a certain map $f:N\to F$ where $F$ is another manifold; We define the *pull-back* $\phi^* f$ of $f$ by

$$\phi^* f = f\phi.$$

So the pull-back of $f$ is a map from $M$ to $F$.

### 8.8.4. Hamiltonian mechanics

Let $M$ be a manifold and $T^*M$ its cotangent bundle. On $T^*M$ 'lives' a canonical symplectic form $\omega$. That means that $\omega$ is a two-form, anti-symmetric and nondegenerate. There are local coordinates $(q,p)$ such that $\omega$ looks like

$$\omega = \sum_{i=1}^{n} dq_i dp_i$$

This can also be considered as the definition of $\omega$, especially in the case $M=\mathbf{R}^n$. Every $H:T^*M\to\mathbf{R}$ defines a vectorfield $X_H$ on $T^*M$ by the relation

$$\iota_{X_H}\omega = dH$$

(There is considerable confusion in the literature due to a sign choice in $\omega$. One should always be very careful with the application of formulas, and check to see whether this choice has been made in a consistent way).
The vectorfield $X_H$ is called the *Hamilton-equation,* and $H$ the *Hamiltonian.* In local coordinates this looks like:

$$\iota_{X_H}\omega = \iota_{\sum_{i=1}^{n}(X_{q_i}\frac{\partial}{\partial q_i} + X_{p_i}\frac{\partial}{\partial p_i})} \sum_{j=1}^{n} dq_j dp_j$$

$$= \sum_{i=1}^{n}(X_{q_i}dp_i - X_{p_i}dq_i).$$

$$dH = \sum_{i=1}^{n}(\frac{\partial H}{\partial q_i}dq_i + \frac{\partial H}{\partial p_i}dp_i).$$

or

$$\dot{q}_i = X_{q_i} = \frac{\partial H}{\partial p_i},$$

$$\dot{p}_i = X_{p_i} = -\frac{\partial H}{\partial q_i}.$$

Let $(T^*M,\omega_1)$ and $(T^*N,\omega_2)$ be two symplectic manifolds and $\phi$ a diffeomorphism between them. We say that $\phi$ is symplectic if $\phi^*\omega_1 = \omega_2$.

Symplectic diffeomorphisms leave the *Hamilton*-equation invariant:

$$\iota_{X_{\phi^* H}}\omega_2 = d\phi^* H = \phi^* \iota_{X_H}\omega_1 = \iota_{\phi^* X_H}\phi^* \omega_1 = \iota_{\phi_* X_H}\omega_2$$

or $X_{\phi^* H} = \phi^* X_H$. We used some results here to be found in (Abr78a).
Let $x_o \in T^* M$. We say that $x_o$ is an *equilibrium point* of $H$ if $dH(x_o)=0$. We call $dimM$ the number of *degrees of freedom* of the system $X_H$. We say that a function or a differential form $\alpha$ is an *integral of motion* for the vectorfield $X$ if

$$L_X \alpha = 0$$

where $L_X$ is defined as

$$L_X \alpha = \iota_X(d\alpha) + (d\iota_X)\alpha.$$

If $\alpha$ is a function, this reduces to

$$L_X \alpha = \iota_X(d\alpha).$$

$H$ itself is an integral of motion of $X_H$, since

$$L_{X_H} = \iota_{X_H} dH = \iota_{X_H}\iota_{X_H}\omega = 0$$

($\omega$ is antisymmetric).
The *Poisson bracket* $\{\ ,\ \}$ is defined on functions (on the cotangent bundle) as follows:

$$\{F,G\} = -\iota_{X_F}\iota_{X_G}\omega.$$

Since $\omega$ is antisymmetric, $\{F,G\} = -\{G,F\}$. Note that

$$\{F,G\} = -\iota_{X_F} dG = -L_{X_F} G$$

and therefore $G$ is invariant with respect to $F$ iff $\{F,G\}=0$. We call $M$ the configuration space and $p_m \in T^*_m M$ is called the *momentum*. If $M$ is the torus, we call the local coordinates *action-angle* variables. Action refers to the momentum and angle to the configuration space coordinates.

## 8.8.5. Morse theory

In considering critical points $x_o$ of a function $f$ we usually consider the *Taylor* expansion of the function in $x_o$. In particular consider the $C^\infty$-function $f:\mathbf{R}^n \to \mathbf{R}$ with $\nabla f(x_o)=0$. The point $x_o$ is called a *non-degenerate critical point* of the function $f$ if the determinant of the second derivative, the *Hessian*, does not vanish at $x_o$. If all critical points of a function are nondegenerate, we call this function a *Morse*-function. We have the following

### 8.8.5.1. Lemma (*Morse*)

Take local coordinates $x_1, \cdots, x_n$ around the critical point $x_o$. Then there exists a local diffeomorphism of $\mathbf{R}^n$ into itself such that the induced function $f^*$ (i.e. $f$ written in the new coordinates given by the local diffeomorphism) can be written as

$$f^*(y) = f(0) - y_1^2 - \cdots - y_k^2 + y_{k+1}^2 + \cdots + y_n^2$$

The number $k$ is called the *index* of $f$ at $x_o$ and it is an invariant of the function. If the index is zero, or $n$, the level curves of $f$ are diffeomorphic to $S^{n-1}$.

*Morse* theory can be useful in studying differential equations if the system can be characterized to some extent by functions as described above, for example integrals of motion. However one should resist the temptation to transform the *Hamiltonian* to a quadratic form, since the local diffeomorphism will, in general, not be symplectic, and therefore one can not deduce the equations of motion from this *Hamiltonian*, at least not with the usual symplectic form.

**References**

Abr78a. Abraham,R. and Marsden,J.E., *Foundations of Mechanics (2nd edition),* The Benjamin/Cummings Publ. Co., Reading, Mass. (1978).

Abr83a. Abraham,R., Marsden,J.E., and Ratiu,T., *Manifolds, Tensor Analysis and Applications,* Addison-Wesley Publ. Co., London (1983).

Br73a. Bröcker,T. and Jänich,K., *Einführung in die Differentialtopologie,* Springer-Verlag, Berlin (1973).

# Index